100道
靚餸
YUMMY
DELICACIES

自序

喚醒味蕾的記憶

食物總能觸動我們的味蕾記憶。在我們的記憶中，最令人難以忘懷的味道莫過於媽媽從廚房端出來一碟碟香噴噴、熱騰騰的菜，因為媽媽把誠意、愛心全都融入食材上。只要她隨性揮動鍋鏟，飢腸轆轆的我們便嘗到生活中最原味平凡、最幸福溫暖、最輕鬆自在的況味——一家人圍在餐桌前品嘗家常便飯。

可是，很忙、很累的上班族、為食材費煞思量的新手、笨手笨腳不諳廚藝的人，總是這樣那樣的理由，對家常飯菜的感情只停留在思念和回憶中，很少動手做菜與家人分享。其實能為家人煮菜是快樂幸福的啊！

為了喚醒大家味蕾的記憶，繼《菲／印傭入廚手記：4材料開飯小菜》後，出版這本新食譜，以平凡尋常的材料，炮製出一百款家常創意的菜式。

期待你或傭人依着一個個簡單明快好學的菜單，而漸漸喜歡做菜，反饋出美味的佳餚，和身邊的親朋好友們一同分享，讓吃菜的人感受到你如陽光般的熱情！

微不足道的小事

上一輩的人常教誨我們「技不壓身」，要對周遭的事或物產生興趣。畢竟多點情趣愛好會活得更充實。能夠讓生命更加充實的技藝，就是美好的事物，美好的事物是要傳遞出去的！

感謝天父，承蒙圓方出版社的邀請及簡小姐的協助，看到出版的第二本食譜，我真的很開心！

德蘭修女 (Mother Teresa) 曾說過：「我們成就不了甚麼大事，卻可以心懷無限的愛去做生活中每一件最小的事。」我雖然平常很忙，做家務、為家人做飯、回中心教授烹飪及陪月課程、到教會參加聚會、與好友逛街⋯⋯卻能夠完成這本書，因為我樂意去做一些微不足道的小事──喜歡做菜，喜歡跟別人分享做菜心得。

不過，使我一股勁的勇於嘗試，瞎子摸路般走一步算一步，那動力源自愛護我的家人──大兒子柏年、媳婦 Polly 和小兒子冠年、廚藝班及陪月班的學生、教會的弟兄姊妹。他們的關心鼓勵與陪伴，令我完成了這一件「小事」！

我謹以此書向天父獻上感恩，也藉此紀念在天父懷抱裏的先夫葉志超先生！

胡影儀

不藏私必學料理

好友共聚

巧用佐料，
提升味道！

Use condiments skillfully to lift the flavour!

薑汁 Ginger juice

鮮榨薑汁可醃製魚或肉類，辟除腥味及膻味，又或用於調味。薑汁可保留一至兩日。

The juice squeezed from fresh ginger can be used to marinate fish or meat to remove its fishy or muttony smell; or as a seasoning for various dishes. The ginger juice can be kept for one to two days.

肉桂 Cinnamon

常見香料之一，味香濃、馥郁，
是中式滷水汁之主要香料；肉桂
粉又可製成甜點、麵包等，用途
廣泛。

Having a strong fragrance, cinnamon is a common spice and also a major spice of
Chinese marinade. Ground cinnamon has a wide usage, such as making desserts and
bread.

南乳 Red fermented beancurd

以大豆、紅麴米及紹酒發酵而成，帶豆香、酒香及甜味，是南乳齋、燜豬手、煮扣肉及南
乳吊燒雞之調味料。

Red fermented beancurd is produced by fermenting soybeans, red yeast rice and Shaoxing wine. It
tastes sweet with a soybean and wine flavour. It is a seasoning for braised vegetables, simmered pork
knuckle, braised pork, and roasted chicken.

金華火腿 Jinhua ham

味甘香、鮮濃，已飛水的火腿肉可炮製蒸燜煮菜式；火腿骨則熬成上湯，作為燉湯、燜鮑魚海味等之佐料，增添鮮香之味。宜選肉色紅潤之原塊金華火腿。

Tasting sweet and smelling fantastic, the blanched meat of Jinhua ham can be applied to steamed or stewed dishes. Its bone can be slowly cooked into stock, serving as a condiment to enhance the flavour of double-steamed soups, braised abalones or dried seafood, and so on. It is ideal to pick the whole pieces of Jinhua ham with ruddy meat.

大地魚 Dried plaice

台灣稱為方魚。將大地魚頭及骨烤香，再煮成上湯，鮮甜惹味；大地魚肉可直接炒煮，或剁成茸製成雲吞餡及撈麵之材料，魚香味濃郁。

The roasted head and bone of a dried plaice can be cooked into a sweet, sensational stock. Its meat can be directly stir-fried or boiled, or finely chopped as an ingredient of wonton filling and mixed noodles. It gives the dish an intense fish flavour.

將沙薑研磨成粉，味道濃烈、帶香味，多製成沙薑雞、沙薑雞翼或沙薑豬扒；或作為鹽焗雞的蘸汁材料。

Dried sand ginger, after being ground into powder, has a strong flavour and aroma. It is mostly used to cook chicken, chicken wings or pork chops. It is also an ingredient of dipping sauce for salt-roasted chicken.

沙薑 Sand ginger

南薑 Galangal

味甜香、少辛辣，常作為烹調咖喱之材料，帶陣陣南薑香氣，與魚類及海鮮很搭配。

A common ingredient of curry dishes, galangal is sweet-scented and less pungent. Its aroma matches beautifully with fish and seafood.

蒸瑤柱汁 Steamed dried scallop sauce

瑤柱之精華，味道鮮甜、香味濃郁，可作為獻汁上湯，又或用於湯羹、蒸水蛋、炒飯等之調味，餸菜提升味道。

he sauce incorporates the essence of dried scallops, which is very aromatic and sweet. It can be used as a stock make thickening sauce, or a seasoning to lift the flavour of soups, steamed eggs, fried rice, and much more.

蒸冬菇，方便烹調！ Steam dried black mushrooms, it is easy!

1. 冬菇用水浸軟，去蒂，冬菇水留用。

2. 冬菇、調味料（糖、紹酒、油、生抽各 1 茶匙）及冬菇水 3 湯匙拌勻，蒸約 15 分鐘，待涼，可儲存備用。

1. Soak dried black mushrooms in water until soft. Remove the stalks and keep the mushroom water.

2. Mix the black mushrooms, seasoning (1 tsp each of sugar, Shaoxing wine, oil and light soy sauce) and 3 tbsps of the mushroom water. Steam for about 15 minutes. When they cool down, store them until it is ready to cook.

浸洗柚皮，去青澀味！ Boil and rinse pomelo skin repeatedly to remove bitterness!

即時睇片

1. 將青色外皮徹底刨去，柚皮放入沸水內蓋面煮約 10 分鐘（不要沾上油），取出，置於水喉下沖水，用手輕輕擠壓水分（共擠壓 3 次）。

2. 再燒滾一鍋水，放入柚皮煮片刻，沖水及擠壓水分。

3. 重複以上步驟 1 次，可去掉青澀味。

1. Peel the green skin entirely from the pomelo. Boil the pomelo skin for about 10 minutes. Make sure that the water must be cover the pomelo and it is not tainted with oil. Then take out the pomelo skin and rinse under running tap water. Gently press the pomelo skin by hand to release water and repeat this action for 3 times.

2. Bring another pot of boiling water. Put in the pomelo skin and cook for a while. Rinse the pomelo skin under running tap water. Gently press them by hand again.

3. Repeat the above steps once more to remove its bitterness.

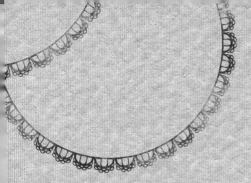

洗豬肚，去異味！ Wash pork stomach to remove odd smell!

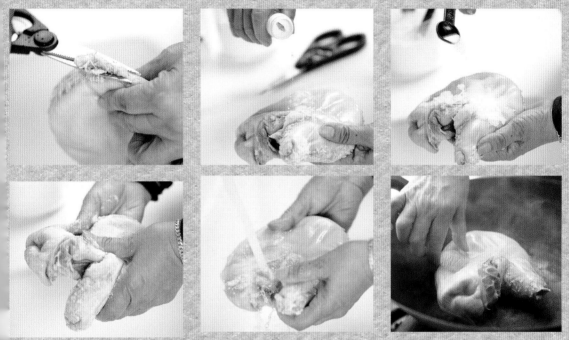

1. 用剪刀去掉豬肚的外邊及肥脂。
2. 於豬肚內外灑上鹽、生粉及醋各1湯匙，用手擦淨，至黏液去掉，用水沖洗，重複以上步驟1次。
3. 燒熱白鑊，放入豬肚略煎兩面。
4. 去除黃色內膜，放入沸水內，加入1茶匙白醋飛水，洗淨即可。

Cut away the edges and fat of the pork stomach with scissors.

Sprinkle 1 tbsp each of salt, caltrop starch and white vinegar on the pork stomach, both inside and outside. Rub off the mucus by hand and then rinse it under running tap water. Repeat this step once more.

Heat a dry wok and then slightly fry both sides of the pork stomach.

Remove the yellow membrane inside the pork stomach. Put it in a pot of water added with 1 tsp of white vinegar and then blanch for a while. Wash it and the preparation is done.

爽彈蝦膠，煮餸多元化！ Make spongy minced shrimp for multiple cooking!

1. 海蝦去殼、去腸，下生粉及鹽各1茶匙略醃，待5分鐘，洗淨。
2. 用乾布包裹蝦肉，徹底壓乾水分，重複2次，至蝦肉帶黏性。
3. 每次取少量蝦肉，用刀面拍成蝦茸。
4. 用刀背輕剁蝦茸（切勿使用刀鋒，會切斷蝦肉纖維）。
5. 加入調味料（蛋白1湯匙、生粉2茶匙、鹽半茶匙、胡椒粉及麻油各少許），順一方向拌勻，見略帶黏性，再撻至起膠即成。

即時睇片 ○······▶

1. Remove the shell and vein of marine shrimps. Marinate with 1 tsp each of caltrop starch and salt and rest for 5 minutes. Then wash the shrimps.

2. Wrap the shrimps in dry cloth and then press it to dry. Repeat the step for 2 times to make them sticky.

3. Bash a small amount of the shrimps each time with the broad side of a knife.

4. Chop the shrimp puree gently with the spine of the knife. Don't use the blade edge as it will cut off the shrimp fibers.

5. Add the seasoning (1 tbsp of egg whites, 2 tsps of caltrop starch, 1/2 tsp of salt, and a little ground white pepper and sesame oil). Stir in one direction until it is a bit sticky. Then throw into the bowl repeatedly to make it gluey.

自家製，香滑芋茸！ Homemade smooth taro puree!

材料 Ingredients

芋頭 1 斤、澄麵 4 湯匙、生油 3 湯匙
600 g taro, 4 tbsps Tang flour, 3 tbsps oil

做法 Method

1. 芋頭切成薄片，蒸約 15 至 20 分鐘，取出，用刀壓成芋茸。
2. 煮滾水 85 毫升，加入澄麵快速拌勻，熄火，加蓋焗片刻，即成熟澄麵。
3. 芋茸、熟澄麵、生油、鹽 1 茶匙及胡椒粉拌勻，搓成幼滑芋欄，即可使用。

1. Finely slice the taro. Steam for about 15 to 20 minutes. Mash into puree with the knife.
2. Bring 85 ml of water to the boil. Add the Tang flour and stir swiftly. Turn off the heat. Put a lid on and rest for a while. The cooked Tang flour is done.
3. Combine the taro, cooked Tang flour, oil, 1 tsp of salt and ground white pepper together. Knead thoroughly into smooth taro dough. It is ready to be used.

在家洗魚腸，easy job！
Wash fish intestine at home, easy job!

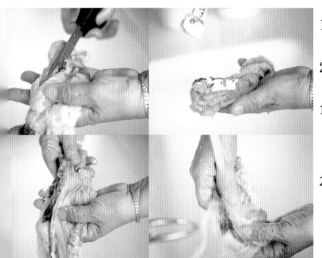

1. 用剪刀去掉鯇魚腸肥膏，去掉膽囊，由末端沿上輕輕剪開魚腸。
2. 用適量生粉、醋及鹽擦淨，放水喉下沖洗，飛水，瀝乾水分備用。

1. Cut away the fat of the grass carp intestine with scissors. Remove the gall bladder. Starting from the end, run the scissors upwards and gently cut open the intestine.
2. Rub the intestine with some caltrop starch, white vinegar and salt. Rinse under running tap water. Blanch for a while and then drain. Ready to use.

家鄉蓮藕餅
Fried Lotus Root Patties

小秘訣 Tips

蓮藕茸與麵粉拌勻，可減少水份，而且材料容易黏合，不易散碎。

Combining the mashed lotus root with flour can reduce the moisture and make the ingredients easily stick together.

材料
Ingredients

蓮藕 10 兩
鯪魚肉 4 兩
豬肉 3 兩
蝦米 3 湯匙
臘肉 3 湯匙（蒸軟、剁幼）
芫茜及葱各少許

375 g lotus root
150 g mud carp meat
113 g pork
3 tbsps dried shrimps
3 tbsps preserved pork
(steamed to soften; finely
chopped)
coriander
spring onion

醃料
Marinade

鹽半茶匙
蛋半個

1/2 tsp salt
1/2 egg

調味料
Seasoning

麵粉 2 湯匙
糖半茶匙
鹽半茶匙
胡椒粉少許

2 tbsps flour
1/2 tsp sugar
1/2 tsp salt
ground white pepper

做法
Method

1. 蓮藕洗淨污泥，去皮，磨成粗茸，加調味料拌勻。
2. 鯪魚肉、豬肉及蝦米剁幼，加入醃料順方向拌至起膠，下蓮藕茸、臘肉、芫茜及葱拌勻。
3. 燒熱鑊，下少許油，用小匙將蓮藕茸放入鑊內，以慢火煎至兩面金黃色即可。

1. Wash away the mud on the lotus root. Remove the skin. Grate into thick puree. Mix well with the seasoning.

2. Finely chop the mud carp meat, pork and dried shrimps. Add the marinade and stir in one direction until sticky. Put in the grated lotus root, preserved pork, coriander and spring onions. Give a good stir.

3. Heat a wok. Add a little oil. Put in the lotus root with a small spoon. Fry over low heat until both sides are golden. Serve.

入廚必修課：煎

技巧重點

- 宜用平底易潔鍋，不容易燒焦，而且用油量少，清洗容易。
- 煎肉扒時，宜用坑紋平底煎鍋，容易將多餘的油脂分隔開來。
- 鍋內的食油要平均分佈，以免食材燒焦。
- 密切注意烹調過程中，食油是否足夠。
- 先用中火，後調中小火煎至金黃色。
- 煎好一面後才翻轉煎另一面，別翻來覆去，可保持美觀的賣相。

Key techniques

- A nonstick pan is ideal for frying food as it will not easily get charred. As only a sm
  amount of oil is needed, it is easy to be cleaned up.

- Using a grill pan for frying steaks will help separate excess grease from the meat easil

- Make sure that the pan is evenly spread with oil to prevent the ingredients fro
  scorching.

- Keep an eye on the frying process to ensure there is enough oil.

- First use medium heat, and then adjust to medium-low to fry the food until golden.

- To keep a good presentation, fry one side of the food until done and then turn it over
  fry the other side. Don't flip it over repeatedly!

火腩大鱔煲
Roast Pork and Japanese Eel Casserole

材料 Ingredients

大鱔半條（1 斤 4 兩）
冬菇 5 隻（處理方法參考 p.12）
燒腩 6 兩
陳皮 1/4 個
薑 4 片
蒜肉 10 粒
芫茜及葱段各少許

1/2 pc Japanese eel (750 g)
5 dried black mushrooms
 (refer to p.12 for method)
225 g roast pork
1/4 dried tangerine peel
4 slices ginger
10 cloves skinned garlic
coriander
sectioned spring onion

醃料 Marinade

鹽 1/4 茶匙
生粉 1 湯匙
胡椒粉少許

1/4 tsp salt
1 tbsp caltrop starch
ground white pepper

調味料 Seasoning

生抽 2 茶匙
蠔油 1 湯匙
糖 1/4 茶匙
水 4 安士（約 115 毫升）
麻油及胡椒粉各少許

2 tsps light soy sauce
1 tbsp oyster sauce
1/4 tsp sugar
4 oz. (115 ml) water
sesame oil
ground white pepper

獻汁 Thickening glaze

生粉 1 1/2 茶匙
水 2 湯匙
老抽 1 茶匙
* 拌勻

1 1/2 tsps caltrop starch
 2 tbsps water
1 tsp dark soy sauce
* mixed well

小秘訣 Tips

燒滾熱油，放入大鱔泡油，令鱔肉嫩滑可口。

Put the Japanese eel in scorching oil to deep-fry for a moment. It makes the meat soft and yum!

做法
Method

1. 陳皮浸軟，刮內瓤，切絲；燒腩切件。
2. 大鱔飛水，過冷河，用刀刮去潺液，切一厘米長段，下醃料拌勻，泡油備用。
3. 燒熱鑊，下凍油，放入蒜肉炸至金黃色，瀝乾油分。
4. 下油 2 湯匙，下薑片、炸蒜肉及冬菇略炒，灒酒，加入燒腩、陳皮絲及調味料煮滾，下大鱔燜至軟脸，埋獻，最後放上芫茜及葱段，原煲上桌。

1. Soak the dried tangerine peel in water to soften. Scrape off the pith. Cut into shreds. Cut the roast pork into pieces.
2. Blanch the Japanese eel. Rinse in cold water. Scrape the skin mucus off with a knife. Cut into sections of 1-cm long. Mix with the marinade. Deep-fry in oil for a moment. Set aside.
3. Heat a wok. Add cool oil. Deep-fry the garlic until golden. Drain.
4. Put in 2 tbsps of oil. Slightly stir-fry the ginger, deep-fried garlic and black mushrooms. Sprinkle with wine. Add the roast pork, tangerine peel and seasoning. Bring to the boil. Put in the Japanese eel and stew until tender. Mix in the thickening sauce. Finally put the coriander and spring onion on top. Serve with the casserole.

香檸辣汁蝦

Prawns in Spicy Lemon Sauce

材料
Ingredients

中蝦 1 斤
紅辣椒 1 隻（切粒）
乾葱 2 粒（切粒）
葱 2 條（切絲）
蒜肉 1 粒（剁茸）
薑茸 1 茶匙

600 g medium prawns
1 red chilli (diced)
2 shallots (diced)
2 sprigs spring onion (shredded)
1 clove skinned garlic (grated)
1 tsp finely chopped ginger

調味料
Seasoning

鹽 1/4 茶匙
糖 2 1/2 茶匙
茄汁 2 湯匙
OK 汁 1 湯匙
檸檬汁 1 湯匙
檸檬皮茸 1 茶匙
胡椒粉少許

1/4 tsp salt
2 1/2 tsps sugar
2 tbsps ketchup
1 tbsp OK sauce
1 tbsp lemon juice
1 tsp grated lemon zest
ground white pepper

獻汁
Thickening glaze

生粉 1/4 茶匙
水 3 湯匙
* 拌勻
1/4 tsp caltrop starch
3 tbsps water
* mixed well

做法
Method

1. 中蝦剪去鬚腳、去腸，洗淨，吸乾水分。
2. 燒熱滾油，放入蝦泡油約 2 分鐘至熟，隔油備用。
3. 燒熱油 2 湯匙，爆香蒜茸、薑茸及乾葱，放入蝦，潷酒，下調味料及紅椒粒，埋獻拌勻上碟，灑上葱絲即可。

1. Cut away the legs and tentacles of the medium prawns with scissors. Devein. Rinse and wipe dry.
2. Heat oil until scorching. Deep-fry the prawns for about 2 minutes, or until done. Drain.
3. Heat 2 tbsps of oil. Stir-fry the garlic, ginger and shallots. Put in the prawns. Sprinkle with wine. Add the seasoning and red chilli. Mix in the thickening glaze. Put on a plate. Sprinkle with the spring onion. Serve.

鮮湯浸柚皮 *Pomelo Skin in Soup*

小秘訣 Tips

- 必須選用中國的沙田柚皮，泰國柚皮不能做成此餚。

- 煮柚皮的鍋別沾上油，否則柚皮容易散爛。

- Only the Chinese pomelo skin can be used for making this dish. Thai pomelo skin is not suitable.

- The pot used for cooking pomelo skin should not be spotted with oil; otherwise the skin will easily fall apart.

材料
Ingredients

柚皮 1 個（處理方法參考 p.12）
鯪魚骨半斤
豬油或生油 1 湯匙
薑 4 片
蒜茸 1 茶匙
滾水 6 杯

1 pomelo skin (refer to p.12 for method)
300 g mud carp bones
1 tbsp lard or oil
4 slices ginger
1 tsp finely chopped garlic
6 cups boiling water

調味料
Seasoning

鹽半茶匙
魚露 2 湯匙
老抽 1 茶匙
片糖 1/3 片
胡椒粉及麻油各少許

1/2 tsp salt
2 tbsps fish sauce
1 tsp dark soy sauce
1/3 slab sugar
ground white pepper
sesame oil

做法
Method

1. 將已處理的柚皮每件切成兩塊，備用。
2. 燒熱鑊，下少許油及薑 2 片，放入已洗淨的鯪魚骨煎香，潷酒，加入滾水煮成魚湯，隔去魚骨，魚湯留用。
3. 燒熱油 4 湯匙，爆香薑片及蒜茸，傾入魚湯及調味料，潷酒略拌，下柚皮煀煮 10 分鐘，加入豬油（或生油），煮至柚皮入味及軟身即可。

1. Cut each prepared pomelo skin into 2 pieces. Set aside.
2. Heat a wok. Add a little oil and ginger. Fry the cleaned mud carp bones until fragrant. Sprinkle with wine. Add the boiling water. Cook into fish soup. Sieve the soup and keep for later use.
3. Heat 4 tbsps of oil. Stir-fry the ginger and garlic until aromatic. Pour in the fish soup and seasoning. Sprinkle with wine. Roughly stir. Put in the pomelo skin and stew for 10 minutes. Add the lard (or oil). Cook the pomelo skin until flavourful and soft. Serve.

家鄉焗鉢仔蟹
Steamed Crab in Earthen Bowl

材料
Ingredients

膏蟹或肉蟹 1 斤
梅頭肉 5 兩
洋葱 1 個（切粒）
鹹蛋黃 2 個（切粒）
雞蛋 1 個
芫茜及葱 2 湯匙（切碎）
蒜茸 1 湯匙

600 g female or male mud crab
188 g pork collar-butt
1 onion (diced)
2 salted egg yolks (diced)
1 egg
2 tbsps coriander and
spring onion (chopped)
1 tbsp finely chopped garlic

調味料
Seasoning

鹽半茶匙
糖 1/4 茶匙
生粉 1 湯匙
生抽 1 1/2 茶匙
紹酒 1 茶匙
胡椒粉少許
鹹蛋白 1 個

1/2 tsp salt
1/4 tsp sugar
1 tbsp caltrop starch
1 1/2 tsps light soy sauce
1 tsp Shaoxing wine
ground white pepper
1 salted egg white

做法
Method

1. 梅頭肉剁碎，與洋葱、雞蛋、鹹蛋黃、芫茜、葱及調味料拌勻，放入瓦鉢內鋪平。
2. 蟹劏好，洗淨，抹乾水分，排放肉碎上，灑上蒜茸、胡椒粉及油，隔水大火蒸約 15 分鐘，取出，將瓦鉢放於焗爐烘至表面香脆即可。

1. Finely chop the pork collar-butt. Mix with the onion, eggs, salted egg yolks, coriander, spring onion and seasoning evenly. Put into an earthen bowl. Make even.
2. Gut the crab. Wash and wipe dry. Arrange on top of the pork mixture. Sprinkle with the garlic, ground white pepper and oil. Steam over high heat for about 15 minutes. Take out. Put the earthen bowl into the oven and roast until the surface is crunchy. Serve.

小秘訣 Tips

將瓦鉢置於焗爐內，用上下火烘至表面金黃色，或置於爐火上，美味香口。

Place the earthen bowl inside the oven and bake with upper and lower heat until the surface is golden, or heat on the stove. It is palatable.

沙爹春卷
Satay Spring Rolls

材料
Ingredients

春卷皮 1 斤
素肉醬（小）1 罐
西芹 4 兩
木耳 4 湯匙
韭黃 2 兩
紅蘿蔔半個
冬菇 6 個（處理方法參考 p.12）
素火腿 3 兩
沙爹醬 2 湯匙
蒜茸 1 茶匙

600 g spring roll pastry
1 small can vegetarian meat paste
150 g celery
4 tbsps wood ear fungus
75 g yellow chives
1/2 carrot
6 dried black mushrooms
(refer to p.12 for method)
113 g vegetarian ham
2 tbsps satay sauce
1 tsp finely chopped garlic

獻汁
Thickening glaze

生粉 1 茶匙
水 1 湯匙
＊拌勻

1 tsp caltrop starch
1 tbsp water
＊mixed well

調味料
Seasoning

糖 1 茶匙
生粉 1 湯匙
生抽 1 茶匙
鹽 1 茶匙
水 3 湯匙
麻油及胡椒粉各少許

1 tsp sugar
1 tbsp caltrop starch
1 tsp light soy sauce
1 tsp salt
3 tbsps water
sesame oil and ground white pepper

做法
Method

1. 西芹及韭黃分別洗淨，切幼條。
2. 木耳浸軟，剪去硬蒂，與冬菇、紅蘿蔔及素火腿分別切成幼條。
3. 燒熱鑊，下蒜茸及沙爹醬炒香，加入其餘材料及調味料炒勻，埋獻，待涼。
4. 春卷皮鋪平，包入餡料，摺成長條狀，下油鑊炸至金黃色，瀝乾油分，即可享用。

1. Rinse and finely shred the celery and yellow chives separately.
2. Soak the wood ear fungus until soft. Cut off the hard stalks. Finely shred wood ear fungus, mushroom, carrot and ham separately.
3. Heat a wok. Stir-fry the garlic and satay sauce until sweet-scented. Add the rest ingredients and seasoning. Stir-fry evenly. Mix in the thickening glaze. Let it cool down.
4. Lay the spring roll pastry flat. Wrap the stuffing. Fold into a bar shape. Deep-fry in oil until golden. Serve.

胡椒豬肚湯

Pork Stomach Soup with White Peppercorns

材料 Ingredients

豬肚 1 個（處理方法參考 p.13)
銀杏 3 兩
白胡椒粒 1 湯匙
豬骨 1 斤
腐皮 3 張
陳皮半個

1 pork stomach (refer to p.13 for method)
113 g gingkoes
1 tbsp white peppercorns
600 g pork bone
3 sheets beancurd skin
1/2 dried tangerine peel

調味料 Seasoning

鹽適量

salt

做法 Method

1. 燒滾水，放入已處理的豬肚及白醋 1 湯匙，飛水，過冷河。
2. 豬骨飛水，過冷河備用。
3. 燒滾清水，加入各材料用中火煲約 2 小時，下鹽調味，取出豬肚切片，伴湯享用。

1. Bring water to the boil. Put in the pork stomach and 1 tbsp of white vinegar. Blanch and rinse with cold water.
2. Scald pork bone. Rinse with cold water and set aside.
3. Bring water to the boil. Add all the ingredients. Cook over medium heat for about 2 hours. Season with salt. Take out the pork stomach and slice. Serve with the soup.

小秘訣 Tips

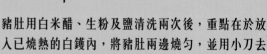

豬肚用白米醋、生粉及鹽清洗兩次後，重點在於放入已燒熱的白鑊內，將豬肚兩邊燒勻，並用小刀去除黃色內膜，以去掉豬肚之異味。。

After washing the pork stomach with white vinegar, caltrop starch and salt twice, you must pay more attention in the following steps. Heat a dry wok and slightly fry both sides of the pork stomach. Then remove the yellow membrane inside the pork stomach. It will remove the odd smell.

越南炸蝦餅
Vietnam Style Deep-fried Shrimp Cakes

小秘訣 Tips

生粉加入蛋白後，要徹底拌勻，以免結成粉粒狀，或用密孔隔篩過濾。

Mix the egg white and caltrop starch thoroughly to avoid forming lumps. Or filter the egg white batter with a fine mesh sieve.

材料
Ingredients

海蝦 1 斤
甜酸雞醬 4 湯匙

600 g marine shrimps
4 tbsps sweet and sour sauce for chicken

蛋白糊
Egg white batter

蛋白 2 個
生粉 2 湯匙

2 egg whites
2 tbsps caltrop starch

蝦調味料
Seasoning for shrimp

蛋白 1 湯匙
鹽 1/3 茶匙
生粉 1 湯匙
胡椒粉及麻油各少許

1 tbsp egg white
1/3 tsp salt
1 tbsp caltrop starch
ground white pepper
sesame oil

做法
Method

1. 蛋白拂勻，加入生粉 2 湯匙拌成蛋白糊。

2. 蝦肉用鹽及生粉各少許洗淨，抹乾水分，用刀剁幼，加入蝦調味料攪拌至起膠，搓成薄圓餅形狀，沾上蛋白糊，放入滾油內炸至金黃熟透，蘸雞醬享用。

1. Whisk the egg white. Add 2 tbsps of caltrop starch. Mix evenly into egg white batter.

2. Clean the shelled shrimps with a little salt and caltrop starch. Rinse and wipe dry. Finely chop. Add the seasoning and stir until gluey. Knead into thin and round cakes. Coat with the egg white batter. Deep-fry in scorching oil until golden and fully cooked. Serve with the sweet and sour dipping sauce.

鍋貼石斑塊

Deep-fried Grouper Dumplings

材料 Ingredients

石斑肉 1 斤
去皮白麵包 4 片
金華火腿茸 2 湯匙
芫茜 1 棵

600 g grouper meat
4 pcs sandwich bread
(brown edges removed)
2 tbsps finely chopped Jinhua ham
1 stalk coriander

醃料 Marinade

鹽及糖各半茶匙
生粉 1 茶匙
胡椒粉少許

1/2 tsp salt
1/2 tsp sugar
1 tsp caltrop starch
ground white pepper

麵糊料 Batter

鹽半茶匙
發粉半茶匙
麵粉 4 安士 （約 115 克）
生粉 3 安士 （約 85 克）
水 4 安士 （約 115 毫升）

1/2 tsp salt
1/2 tsp baking powder
4 oz. (115 g) flour
3 oz. (85 g) caltrop starch
4 oz. (115 ml) water

小秘訣 Tips

麵包容易搶火焦燶，注意火候不要調至太大，以中大火即可。

Use medium-high heat for deep-frying as the sandwich bread will easily burnt over high heat.

做法
Method

1. 麵包切成小塊，備用。
2. 石斑肉洗淨，抹乾水分，切成麵包相同大小，下醃料拌勻。
3. 麵糊料調勻，最後加入油 4 湯匙在麵糊面，放入雪櫃冷藏半小時。
4. 麵糊調勻，將每件石斑肉沾上麵糊，鋪於麵包上，灑少許火腿茸及芫茜葉。
5. 燒熱油，放入石斑鍋貼（面向下）炸至金黃色即成。

1. Cut the sandwich bread into small pieces. Set aside.
2. Rinse the grouper meat. Wipe dry. Cut into the size same as the sandwich bread. Mix with the marinade evenly.
3. Combine the ingredients for batter. Mix well. Add 4 tbsps of oil on the surface of the batter. Chill for 1/2 hour.
4. Mix the batter. Coat each piece of grouper meat with batter. Lay on the sandwich bread. Sprinkle with a little Jinhua ham and coriander leaves.
5. Heat oil. Deep-fry the grouper dumplings (the grouper is downward) until golden. Serve.

生炒排骨

Sweet and Sour Pork Spareribs

材料
Ingredients

腩排 14 兩
青、紅甜椒各半個
葱白 4 條
菠蘿 2 片
蒜肉 2 粒（剁茸）

525 g pork spareribs
1/2 green bell pepper
1/2 red bell pepper
4 sprigs white part of spring onion
2 slices pineapple
2 cloves garlic (finely chopped)

調味料
Seasoning

鹽半茶匙
茄汁 2 湯匙
糖及白醋各 4 湯匙
橙紅粉 1/8 茶匙
山楂片 5 片（用開水 4 湯匙浸溶）

1/2 tsp salt
2 tbsps ketchup
4 tbsps sugar
4 tbsps white vinegar
1/8 tsp orange red food colour powder
5 haw flakes (dissolved in 4 tbsps drinking water)

醃料
Marinade

鹽及糖各半茶匙
生抽 1 1/2 茶匙
蛋黃 1 個

1/2 tsp salt
1/2 tsp sugar
1 1/2 tsps light soy sauce
1 egg yolk

獻汁
Thickening glaze

生粉 1 茶匙
水 2 湯匙
* 拌勻

1 tsp caltrop starch
2 tbsps water
* mixed well

小秘訣 Tips

• 橙紅粉令餸菜的色澤更亮麗，可隨意加入與否。

• 山楂片的微酸微甜，令整道餸增添不少風味。

• 山楂緊記先用開水浸溶，別整塊加入拌炒，以免難以溶解。

• The orange red food colour powder will give a beautiful glaze on the dish. You can choose to add it or not.

• The light sweet, sour taste of haw flakes adds flavour to the dish.

• Soak the haw flakes in drinking water until they dissolve. Do not stir-fry whole pieces of haw flakes as it can hardly make them dissolve.

做法
Method

1. 腩排洗淨，抹乾水分，加入醃料醃片刻。
2. 青紅甜椒及菠蘿分別切塊，備用。
3. 鑊內燒熱油半熱，下青紅甜椒泡油，盛起。
4. 燒熱油，腩排沾上乾生粉，下油鑊炸至金黃色，盛起，待油再滾，下腩排再炸一次。
5. 燒熱油 2 湯匙，下蒜茸及葱白爆香，加入調味料煮滾，潷酒，下腩排快炒，埋獻，最後拌入甜椒及菠蘿，上碟享用。

1. Rinse the pork spareribs. Wipe dry. Add the marinade and rest for a while.
2. Cut the green bell pepper, red bell pepper and pineapple into pieces separately. Set aside.
3. Heat oil in a wok. When the oil is warm, put in the green and red bell peppers. Deep-fry for a while. Set aside.
4. Heat oil. Coat the spareribs with caltrop starch. Deep-fry until golden. Dish up. When the oil comes to the boil again, deep-fry the spareribs once more.
5. Heat 2 tbsps of oil. Stir-fry the garlic and white part of spring onion until fragrant. Pour in the seasoning. Bring to the boil. Sprinkle with wine. Put in the spareribs and stir-fry swiftly. Mix in the thickening glaze. Finally add the bell peppers and pineapple. Mix evenly. Put on a plate and serve.

黑椒鹹豬手煲
Salted Pork Knuckle with Black Pepper in Casserole

小秘訣 Tips

加入冰糖及白米醋燜豬手，令肉質容易軟化，肉軟酥香！

To stew the pork knuckle with rock sugar and white rice vinegar will soften the meat texture.

材料 Ingredients

鹹豬手 1 隻（約 1 斤）
洋蔥 1 個
蒜茸 1 湯匙
黃芽白半斤

1 salted pork knuckle (about 600 g)
1 onion
1 tbsp finely chopped garlic
300 g Peking cabbage

調味料 Seasoning

黑胡椒粒、老抽及生抽各 1 湯匙
冰糖 2 湯匙
香葉 3 片
白米醋 1 湯匙
麻油少許
水 3 杯

1 tbsp black peppercorns
1 tbsp dark soy sauce
1 tbsp light soy sauce
2 tbsps rock sugar
3 bay leaves
1 tbsp white rice vinegar
sesame oil
3 cups water

做法 Method

1. 洋蔥去外衣，洗淨，切塊；黃芽白洗淨，切塊。
2. 豬手洗淨，放入滾水煮半小時，過冷河備用。
3. 燒熱鑊，下油 2 湯匙，爆香洋蔥、蒜茸及豬手，下調味料燜 50 分鐘，關火焗半小時。
4. 瓦煲內掃上油，下黃芽白及豬手煮片刻，下生粉水埋獻，原煲上桌。

1. Skin the onion. Rinse and cut into pieces. Rinse the Peking cabbage. Cut into pieces.
2. Wash the pork knuckle. Cook in boiling water for 1/2 hour. Rinse in cold water. Set aside.
3. Heat a wok. Add 2 tbsps of oil. Stir-fry the onion, garlic and pork knuckle until fragrant. Pour in the seasoning and simmer for 50 minutes. Turn off heat. Leave for 1/2 hour with the lid on.
4. Spread oil inside a casserole. Put in the Peking cabbage and pork knuckle. Cook for a while. Mix in the caltrop starch solution. Serve with the casserole.

蒜茸大蝦
Garlic Prawns

材料 Ingredients

大蝦 2 隻
蒜茸 1 湯匙
牛油 2 湯匙

2 prawns
1 tbsp finely chopped garlic
2 tbsps butter

醃料 Marinade

鹽 1/4 茶匙
胡椒粉少許
生粉 1 茶匙

1/4 tsp salt
ground white pepper
1 tsp caltrop starch

做法 Method

1. 大蝦洗淨,抹乾水分,灑上少許胡椒粉及鹽拌勻,撲上生粉,放入滾油內炸熟,盛起。
2. 下牛油 2 湯匙,以慢火爆香蒜茸,放入大蝦拌勻,瀳酒,加蓋略焗(約需半分鐘),原隻上碟享用。

1. Wash the prawns. Wipe dry. Sprinkle with a little ground white pepper and salt. Mix evenly. Coat with caltrop starch. Deep-fry in scorching oil until done. Set aside.
2. Add 2 tbsps of butter. Stir-fry the garlic over low heat until aromatic. Put in the prawns and mix well. Sprinkle with wine. Cook with a lid on for about 1/2 minute. Serve with the whole prawns.

小秘訣 Tips

若不喜歡炒吃,可將大蝦炸熟,置於錫紙上,塗抹少許牛油,灑上蒜茸及少許鹽,放入焗爐用大火焗約 5 分鐘亦可。

If you don't like stir-fried prawns, you may deep-fry the prawns until done. Then put them on aluminum foil; spread a little butter on the prawns; sprinkle with finely chopped garlic and a little salt; and bake in an oven over high heat for about 5 minutes.

蠔仔餅
Fried Oyster Cake

材料 Ingredients

蠔仔 12 兩
鴨蛋 5 個
薯粉 2 湯匙
蔥 3 條
青蒜 2 棵

450 g baby oysters
5 duck eggs
2 tbsps potato starch
3 sprigs spring onion
2 stalks green garlic

調味料 Seasoning

鹽及生抽各 1 茶匙
糖及胡椒粉各半茶匙
薑汁 1 湯匙
麻油少許

1 tsp salt
1 tsp light soy sauce
1/2 tsp sugar
1/2 tsp ground white pepper
1 tbsp ginger juice
sesame oil

小秘訣 Tips

蠔仔拌入薯粉，煎出來的蠔餅更挺身，口感更佳。

Combining the baby oysters with potato starch will make the fried cake stiffer and more delicious.

做法 Method

1. 鴨蛋拂勻；蔥及青蒜洗淨，分別切粒。
2. 蠔仔用生粉 2 湯匙、鹽 1 湯匙及白醋 1 茶匙拌勻，洗擦，用水沖淨，盛於笪箕內，瀝乾水分，下調味料拌勻。
3. 薯粉用水 2 湯匙調勻，加入蠔仔攪拌。
4. 燒熱鑊，下油 6 湯匙，加入青蒜粒爆香，下蠔仔漿弄成薄薄一層，煎至兩面熟透，灑入少許蔥粒及半份蛋汁，煎至蛋呈金黃色，翻轉，再灑入蔥粒及餘下蛋汁，煎至金黃色即可。

1. Whisk the duck eggs. Rinse the spring onion and green garlic. Dice separately.
2. Mix the baby oysters with 2 tbsps of caltrop starch, 1 tbsp of salt and 1 tsp of white vinegar. Rub to clean. Rinse thoroughly. Drain in a strainer. Add the seasoning and mix well.
3. Combine the potato starch with 2 tbsps of water. Add the baby oysters. Give a good stir.
4. Heat a wok. Add 6 tbsps of oil. Stir-fry the green garlic until sweet-scented. Put in the oyster mixture and make into a thin layer. Fry until both sides are fully done. Sprinkle with a little diced spring onion and 1/2 portion of the egg wash. Fry until the egg is golden. Flip over. Sprinkle with the spring onion and pour in the rest egg wash. Fry until golden. Serve.

炒桂花蟹
Stir-fried Crab with Eggs

小秘訣 Tips

蛋汁下鑊後快炒，待蛋汁未凝固即放入蟹件，令滑蛋緊緊地包着蟹件。

Stir-fry the egg mixture swiftly. When it is not yet set, put in the crab pieces so that they can be entirely wrapped in the eggs.

材料 Ingredients

花蟹 1 斤
雞蛋 4 個
瑤柱 3 粒
薑茸半湯匙
蒜茸 1 茶匙
芫茜碎及葱碎 1 湯匙

600 g blue crab
4 eggs
3 dried scallops
1/2 tbsp finely chopped ginger
1 tsp finely chopped garlic
1 tbsp chopped coriander
and spring onion

調味料 Seasoning

生粉 1 1/2 茶匙
生抽 1 茶匙
鹽及糖各 1/4 茶匙
薑汁 1 茶匙
蒸瑤柱汁 4 湯匙

1 1/2 tsps caltrop starch
1 tsp light soy sauce
1/4 tsp salt
1/4 tsp sugar
1 tsp ginger juice
4 tbsps sauce from steamed dried scallop

做法 Method

1. 蟹洗淨，抹乾水分，斬件，下生粉 1 湯匙及胡椒粉少許拌勻，泡油備用。

2. 瑤柱用水浸透，下少許油、薑汁、酒及糖蒸 10 分鐘至熟，拆絲，蒸瑤柱汁留用。

3. 雞蛋拂勻，放入瑤柱絲及調味料拌勻。

4. 燒熱鑊，下油 3 湯匙，加入薑茸及蒜茸爆香，傾入步驟 3 蛋料炒勻，下蟹件炒片刻，潠酒，上碟，灑上芫茜及葱即可。

1. Wash the crab thoroughly. Wipe dry. Chop into pieces. Mix with 1 tbsp of caltrop starch and a little ground white pepper. Deep-fry in oil for a moment. Set aside.

2. Soak the dried scallops in water until soft. Add a little oil, ginger juice, wine and sugar. Steam for 10 minutes, or until done. Tear into shreds. Keep the steamed sauce for later use.

3. Whisk the eggs. Put in the dried scallop shreds and seasoning. Mix well.

4. Heat a wok. Add 3 tbsps of oil. Stir-fry the ginger and garlic until sweet-scented. Pour in the egg mixture from Step 3. Stir-fry evenly. Put in the crab and stir-fry for a moment. Sprinkle with wine. Dish up. Sprinkle with the coriander and spring onion. Serve.

蔗 蝦

Deep-fried Minced Shrimp on Sugar Canes

材料
Ingredients

甘蔗 2 條（約 5 吋長）
蝦膠 12 兩（做法參考 p.14）
馬蹄肉 2 個
肥肉粒 2 湯匙（煮熟）

2 sugar canes (about 5-inch long)
450 g minced shrimp
(refer to p.14 for method)
2 skinned water chestnuts
2 tbsps diced fat pork (cooked)

調味料
Seasoning

蛋白 1 湯匙
鹽半茶匙
麻油及胡椒粉各少許
生粉 1 湯匙

1 tbsp egg white
1/2 tsp salt
sesame oil
ground white pepper
1 tbsp caltrop starch

醬汁
Sauce

魚露 1 湯匙
檸檬汁半湯匙
白醋 1 湯匙
糖 1 湯匙
蒜茸 3/4 湯匙
冷開水 3 湯匙
＊拌勻

1 tbsp fish sauce
1/2 tbsp lemon juice
1 tbsp white vinegar
1 tbsp sugar
3/4 tbsp finely chopped garlic
3 tbsps cold drinking water
* mixed well

小秘訣 Tips

包釀蝦膠時，弄濕雙手，容易將蝦膠抹平。

Damp both hands before wrapping in the shrimp mixture. It is easy to make it even.

做法
Method

1. 馬蹄肉剁成幼粒，備用。
2. 蝦膠、馬蹄粒、肥肉粒及調味料順方向拌勻，再撻至起膠。
3. 甘蔗洗淨，用刀直切，分成兩段。在甘蔗中段抹上少許粟粉，包上蝦膠。
4. 蔗蝦放入滾油內，炸至金黃熟透，瀝乾油分，蘸醬汁或泰式甜酸雞醬享用。

1. Finely chop the water chestnuts. Set aside.
2. Stir the minced shrimp, water chestnuts, diced fat pork and seasoning in one direction. Mix well. Throw into a bowl repeatedly until it is sticky.
3. Rinse the sugar canes. Cut through the sugar cane into 2 sticks. Spread a little cornflour on the middle part of the sugar cane. Wrap in the shrimp mixture.
4. Deep-fry the shrimp sugar cane in scorching oil until golden and fully cooked. Drain. Serve with the dipping sauce or Thai sweet and sour sauce for chicken.

梅子蒸鴨
Steamed Duck with Pickled Plums

材料
Ingredients

光鴨 1 隻（約 2 1/2 斤）
酸梅 4 兩
蒜肉 6 粒

1 duck (about 1.5 kg)
150 g pickled plums
6 cloves skinned garlic

調味料
Seasoning

磨豉醬 1 湯匙
冰糖 5 兩（剁碎）
鹽 1 茶匙

1 tbsp ground bean sauce
188 g rock sugar (chopped up)
1 tsp salt

做法
Method

1. 鴨洗淨，抹乾水分，用老抽 3 湯匙及鹽 1 茶匙均勻地塗抹鴨身，略醃待上色。
2. 酸梅去核，搗碎；蒜肉剁碎，加入調味料，拌成調味醬。
3. 將調味醬釀入鴨腹內，用竹籤或燒烤串緊緊地縫合。燒熱油 3 湯匙，放入鴨煎至金黃色，盛起，瀝乾油分，鴨肚向上隔水蒸 50 分鐘，傾出鴨腹內醬汁留用。
4. 將鴨斬件，上碟，澆上步驟 3 之醬汁，即可享用。

1. Wash the duck thoroughly. Wipe dry. Spread 3 tbsps of dark soy sauce and 1 tsp of salt evenly on the duck. Marinate for a moment to let it colour.
2. Remove the cores of the pickled plums. Mash up. Finely chop the garlic. Add the seasoning. Mix well into paste.
3. Stuff the seasoned paste into the duck cavity. Sew up tightly with a bamboo stick or skewer. Fry in 3 tbsps of hot oil until golden. Drain. Steam for 50 minutes (the duck cavity is upward). Pour out the sauce from the duck cavity. Keep the sauce.
4. Chop the duck into pieces. Put on a plate. Sprinkle with the sauce from Step 3. Serve.

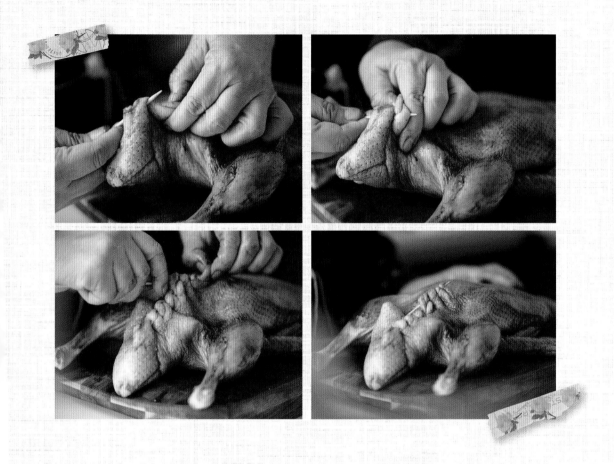

小秘訣 Tips

- 現時市面沒有鮮鴨出售，超市及凍肉店有冰鮮洋鴨，解凍後即可烹調。
- 可用燒烤串或針線緊緊縫起鴨腹，在蒸鴨過程中，讓調味醬慢慢滲入鴨肉。
- 調味料用的冰糖要徹底剁碎，有效散發香味。
- There is no live duck available in the market at present. Chilled Muscovy ducks are sold in supermarkets and frozen food shops. Defrost the duck before cooking.
- You can use a skewer or needle and thread to sew up the duck cavity. It is to let the seasoned paste gradually infuse into the meat during the steaming process.
- The rock sugar for seasoning must be thoroughly chopped up to let the aroma diffuse.

酥炸九肚魚
Deep-fried Bombay Duck
with Spice Salt

材料 Ingredients

九肚魚 1 斤
雞蛋 1 個
淮鹽 1/4 茶匙
紅椒圈適量

600 g Bombay duck
1 egg
1/4 tsp spice salt
red chilli rings

醃料 Marinade

鹽半茶匙
胡椒粉少許

1/2 tsp salt
ground white pepper

做法 Method

1. 九肚魚劏淨，去掉頭尾兩端，每條切成三段，洗淨，抹乾水分。

2. 九肚魚用醃料醃約5分鐘，加入蛋液拌勻，撲上生粉。

3. 燒熱油，下九肚魚炸至金黃色，盛起，瀝乾油分，灑上淮鹽，以紅椒圈裝飾即成。

1. Gill the Bombay duck. Cut away the head and tail. Cut each fish into 3 sections. Rinse and wipe dry.

2. Mix the fish with the marinade and rest for about 5 minutes. Stir in the egg wash. Coat with caltrop starch.

3. Heat oil. Deep-fry the fish until golden and drain. Sprinkle with the spice salt. Decorate with the red chilli rings. Serve.

小秘訣 Tips

- 將油燒熱至大滾，九肚魚才炸得香脆可口。

- 淮鹽可自行在家準備，將鹽1茶匙用慢火炒至微黃，再下五香粉半茶匙炒片刻，可留用。

- The oil used for deep-frying must be very hot so as to make the Bombay duck crunchy.

- You can make spice salt at home. Stir-fry 1 tsp of salt over low heat until it turns light brown. Add 1/2 tsp of five-spice powder and stir-fry for a while. It is done.

芋茸鳳尾蝦
Deep-fried Taro Shrimps

即時睇片 ○······▶

材料
Ingredients

蝦 24 隻
芋茸 1 份（做法參考 p.15）
24 shrimps
1 portion taro puree
(refer to p.15 for method)

小秘訣 Tips

包入蝦時，芋茸要揑得均勻，美觀
之餘，亦容易控制熟度。

Knead the taro puree evenly when
wrapping in a shrimp. It will keep
the nice presentation and is easily
to deep-fry until done.

做法
Method

1. 蝦去殼、去腸，留尾洗淨，抹乾水分，下鹽
   1/4 茶匙、胡椒粉及麻油各少許拌勻，最後撲
   上生粉。

2. 芋茸分成 24 份，取一份芋茸包入蝦（蝦尾在
   外），放入滾油炸至金黃色即可。

1. Shell the shrimps except the tail. Devein and
   rinse. Wipe dry. Add 1/4 tsp of salt, a little
   ground white pepper and sesame oil. Mix well.
   Coat with caltrop starch at last.

2. Divide the taro puree into 24 portions. Wrap a
   shrimp (tail outside) in a portion of the taro
   puree. Deep-fry in hot oil until golden. Serve.

056

豉椒炒鵝腸

Stir-fried Goose Intestine with Fermented Black Beans and Bell Peppers

小秘訣 *Tips*

炒鵝腸時用大火快炒，動作迅速，能吃出爽脆的口感。

Swiftly stir-fry the goose intestine over high heat to make it taste crunchy.

材料 Ingredients

鵝腸 1 斤
青、紅甜椒各 1 個
葱 2 條（切段）
薑 3 片
蒜肉 2 粒（切片）
豆豉 1 湯匙（洗淨，去皮）
白醋適量

600 g goose intestine
1 green bell peppers
1 red bell peppers
2 sprigs spring onion (cut into short sections)
3 slices ginger
2 cloves skinned garlic (sliced)
1 tbsp fermented black beans (rinsed; skinned)
white vinegar

調味料 Seasoning

生抽 2 茶匙
糖 1 茶匙
胡椒粉、麻油及鹽各少許

2 tsps light soy sauce
1 tsp sugar
ground white pepper
sesame oil
salt

獻汁 Thickening glaze

生粉 2 茶匙
清水 2 湯匙
* 拌勻

2 tsps caltrop starch
2 tbsps water
* mixed well

做法 Method

1. 鵝腸用鹽、生粉及白醋各 1 湯匙擦洗，加入少許白醋飛水，盛起，用水沖淨，抹乾水分，切成短度。

2. 甜椒洗淨，去籽，切成角形，用油略炒，瀝乾水分。

3. 燒紅鑊，下油 3 湯匙，加入薑片、蒜片及豆豉略炒，下鵝腸及調味料炒片刻，最後加入青紅甜椒及葱段，潷酒，埋獻即成。

1. Rub the goose intestine with 1 tbsp of salt, 1 tbsp of caltrop starch and 1 tbsp of white vinegar. Blanch with a little white vinegar. Dish up and rinse. Wipe dry. Cut into short sections.

2. Rinse and deseed the bell peppers. Cut into triangles. Roughly stir-fry with oil. Drain.

3. Heat a wok. Add 3 tbsps of oil. Add the ginger, garlic and fermented black beans. Slightly stir-fry. Put in the goose intestine and seasoning. Stir-fry for a while. Finally add the bell peppers and spring onion. Sprinkle with wine. Mix in the thickening glaze. Serve.

薑葱焗生蠔
Oysters with Ginger and Spring Onions

材料
Ingredients

生蠔 1 斤
薑絲 3 湯匙
葱絲 2 兩
蒜肉 1 粒（剁茸）

600 g oysters
3 tbsps shredded ginger
75 g shredded spring onion
1 clove skinned garlic
(finely chopped)

調味料
Seasoning

鹽及糖各 1/4 茶匙
生抽半茶匙
蠔油 1 湯匙
麻油及胡椒粉各少許

1/4 tsp salt
1/4 tsp sugar
1/2 tsp light soy sauce
1 tbsp oyster sauce
sesame oil
ground white pepper

醃料
Marinade

粟粉 1 湯匙
胡椒粉少許

1 tbsp corn flour
ground white pepper

獻汁
Thickening glaze

生粉 1 茶匙
水 1 湯匙
* 拌勻

1 tsp caltrop starch
1 tbsp water
* mixed well

小秘訣 Tips

- 擦洗生蠔時，加點白醋可去除腥味。
- 生蠔飛水時別太久，稍見蠔肉漲滿即可盛起。
- Rubbing the oysters with a little white vinegar will help remove their fishy smell.
- Do not blanch the oysters for too long. When they slightly plump out, dish up.

做法
Method

1. 生蠔用生粉及白醋各1湯匙、粗鹽少許擦淨,用水洗淨。
2. 燒滾水,放入生蠔、葱2條及薑數片,飛水,盛起,抹乾水分,下醃料拌勻,泡油備用。
3. 燒熱油2湯匙,下薑絲、葱絲及蒜茸爆香,潽酒,放入蠔及調味料煮約半分鐘,埋獻即成。

1. Rub the oysters with 1 tbsp of caltrop starch, 1 tbsp of white vinegar and a little coarse salt. Wash thoroughly. Rinse well.

2. Bring water to the boil. Put in the oyster, 2 sprigs of spring onion and a couple of ginger slices. Blanch and drain. Wipe dry. Mix with the marinade. Deep-fry in oil for a while. Set aside.

3. Heat 2 tbsps of oil. Stir-fry the ginger, spring onion and garlic until fragrant. Sprinkle with wine. Put in the oyster and seasoning. Cook for about 1/2 minute. Mix in the thickening glaze Serve.

麻辣雞煲
Spicy Chicken Casserole

材料 Ingredients

雞半隻
三色甜椒各半個
大豆芽 4 兩
薑 4 片
蒜片 1 湯匙
豆瓣醬 1 湯匙
葱 2 條（切短度）

1/2 chicken
1/2 green bell pepper
1/2 red bell pepper
1/2 yellow bell pepper
150 g soy bean sprouts
4 slices ginger
1 tbsp sliced garlic
1 tbsp chilli bean sauce
2 sprigs spring onion
(cut into short sections)

調味料 Seasoning

鹽、糖及老抽各 1 茶匙
白醋 1 茶匙
水 1 杯

1 tsp salt
1 tsp sugar
1 tsp dark soy sauce
1 tsp white vinegar
1 cup water

醃料 Marinade

生抽及老抽各半湯匙
油半湯匙
胡椒粉少許
生粉 1 湯匙（後下）

1/2 tbsp light soy sauce
1/2 tbsp dark soy sauce
1/2 tbsp oil
ground white pepper
1 tbsp caltrop starch (added last)

獻汁 Thickening glaze

生粉 1 茶匙
水 3 湯匙
* 拌勻

1 tsp caltrop starch
3 tbsps water
* mixed well

小秘訣 Tips

- 徹底將雞件抹乾水分，才加入醃料拌勻，待不少於半小時，讓調味香料滲入雞肉。
- 雞件泡油前才灑入生粉拌勻，令雞塊香脆入味。
- The chicken must be dried thoroughly before marinating. Let it rest for at least 1/2 hour so that the seasoning can permeate into the chicken.
- Mix the chicken with caltrop starch right before deep-frying. The chicken will be crisp and flavourful.

做法
Method

1. 三色甜椒去籽，洗淨，切件，用油略泡即盛起。

2. 雞洗淨，抹乾水分，斬件，加醃料醃半小時，下生粉略拌，炸至微黃色備用。

3. 大豆芽洗淨，瀝乾水分，白鑊烘炒至乾，加水 3 湯匙拌炒，轉放瓦煲內。

4. 下油 2 湯匙，下薑片、蒜片、豆瓣醬及雞件爆香，潷酒，下調味料煮片刻，加入三色甜椒拌勻，埋獻，鋪入瓦煲內，放上葱段，澆上麻油 1 湯匙煮滾，原煲上桌。

1. Deseed and rinse bell peppers. Cut into pieces. Slightly deep-fry in oil. Set aside.

2. Wash the chicken. Wipe dry. Chop into pieces. Mix with the marinade and rest for 1/2 hour. Roughly mix in caltrop starch. Deep-fry until light brown. Set aside.

3. Rinse the soy bean sprouts. Drain well. Stir-fry in a wok without oil until dry. Add 3 tbsps of water. Transfer to a casserole.

4. Put in 2 tbsps of oil. Stir-fry the ginger, garlic, chilli bean sauce and chicken until fragrant. Sprinkle with wine. Add the seasoning and cook for a moment. Put in the bell peppers and mix well. Mix in the thickening glaze. Lay on the soy bean sprouts in the casserole. Put the spring onion on top. Drizzle with 1 tbsp of sesame oil. Bring to the boil. Serve with the casserole.

蜜汁燒鱔
Honey Glazed Japanese Eel

白鱔 1 斤
蜜糖 1 湯匙
紹酒 1 湯匙
檸檬汁 1 茶匙

600g Japanese eel
1 tbsp honey
1 tbsp Shaoxing wine
1 tsp lemon juice

叉燒醬 4 茶匙
糖 1 茶匙
鹽半茶匙
胡椒粉及麻油各少許
生粉 2 茶匙

4 tsps BBQ pork sauce
1 tsp sugar
1/2 tsp salt
ground white pepper
sesame oil
2 tsps caltrop starch

做法
Method

1. 白鱔放入熱水內待一會，用刀刮去潺液，洗淨，抹乾水分，起肉，切幼片（別切斷魚皮），下醃料拌勻。

2. 燒熱鑊，下油 4 湯匙，放入白鱔以慢火煎至金黃色，下蜜糖及紹酒拌勻，上碟，最後灑上檸檬汁即成。

1. Put the Japanese eel into hot water. Rest for a while. Scrape the skin mucus off with a knife. Wash thoroughly. Wipe dry. Remove the bone. Cut into fine slice (don't cut through the skin). Mix with the marinade.

2. Heat a wok. Add 4 tbsps of oil. Fry the Japanese eel over low heat until golden. Put in the honey and Shaoxing wine. Mix well. Put on a plate. Drizzle with the lemon juice. Serve.

小秘訣 Tips

白鱔宜放入 60℃ 熱水內待片刻，水溫太熱容易令魚皮脫掉。

Rest the Japanese eel in hot water of 60 ℃ for a moment. The skin will fall off if the water is too hot.

怪味雞
Chicken with Multi-flavours

小秘訣 Tips

想吃爽脆的雞皮,待雞蒸熟或浸熟後,即放入冰水內浸泡,掛起晾乾即可。

If you want the chicken skin crunchy, soak the fully cooked chicken (steamed or soaked in boiled water) immediately in ice water, and then hang to dry.

材料 Ingredients

雞 1 隻
蒜茸、薑茸及葱絲各適量

1 chicken
finely chopped garlic
finely chopped ginger
shredded spring onion

醬汁 Sauce

麻油、芝麻醬及黑醋各 1 湯匙
花椒粉及辣椒油各 1 茶匙
老抽 2 湯匙
上湯半杯

1 tbsp sesame oil
1 tbsp sesame paste
1 tbsp black vinegar
1 tsp ground Sichuan peppercorn
1 tsp chilli oil
2 tbsps dark soy sauce
1/2 cup stock

做法 Method

1. 雞洗淨，用鹽 2 茶匙抹勻雞腔內外，隔水蒸約 25 分鐘至熟透，斬件上碟。
2. 醬汁預先調好拌勻。
3. 燒熱鍋下油，炒香蒜茸及薑茸，下醬汁略煮，放入葱絲拌勻，澆在雞件上即成。

1. Wash the chicken. Evenly spread 2 tsps of salt on the chicken inside and outside. Steam for about 25 minutes, or until fully cooked. Chop into pieces. Put on a plate.
2. Mix the sauce evenly beforehand.
3. Heat a pot. Put in oil. Stir-fry the garlic and ginger until fragrant. Put in the sauce and cook for a while. Mix in the spring onion. Pour on the chicken. Serve.

煎桂花魚肚 *Fried Fish Maw with Eggs*

砂爆魚肚 1 兩
雞蛋（大）5 個
芫茜 1 棵（切碎）
薑 2 片
葱 4 條（取 2 條切粒）

38 g fried fish maw
5 large eggs
1 stalk coriander (chopped up)
2 slices ginger
4 sprigs spring onion
(2 sprigs for dicing)

調味料
Seasoning

鹽 1 茶匙
粟粉 1 湯匙
生抽半茶匙
紹酒 1 茶匙
麻油及胡椒粉各少許

1 tsp salt
1 tbsp corn flour
1/2 tsp light soy sauce
1 tsp Shaoxing wine
sesame oil
ground white pepper

做法
Method

1. 魚肚用清水浸軟（約 1 小時），擠乾水分。
2. 燒熱鑊下少許油，加入薑 2 片及葱 2 條爆香，潷酒，下白醋 1 茶匙及水 2 杯，放入魚肚煮約 5 分鐘，沖洗後擠乾水分，切成幼粒。
3. 燒紅鑊下少許油，放入魚肚粒炒香，潷酒，下調味料炒勻，上碟備用。
4. 雞蛋拂勻，下油半湯匙、鹽 1/4 茶匙及芫茜葱拌勻，最後下魚肚混和。
5. 燒熱油 3 湯匙，舀入蛋漿混合料 2 湯匙，用慢火煎成荷包蛋形狀，上碟享用。

1. Soak the fish maw in water to soften (about 1 hour). Squeeze water out.
2. Heat a wok. Add a little oil. Stir-fry 2 slices of ginger and 2 sprigs of spring onion until fragrant. Sprinkle with wine. Add 1 tsp of white vinegar and 2 cups of water. Put in the fish maw and cook for about 5 minutes. Rinse it and squeeze water out. Finely diced.
3. Heat a wok. Add a little oil. Stir-fry the fish maw until sweet-scented. Sprinkle with wine. Put in the seasoning and stir-fry evenly. Set aside.
4. Whisk the eggs. Add 1/2 tbsp of oil, 1/4 tsp of salt, coriander and spring onion. Mix well. Finally mix in the fish maw.
5. Heat 3 tbsps of oil. Put in 2 tbsps of the egg mixture. Fry over low heat into a round shape. Put on a plate. Serve.

小秘訣 Tips

砂爆魚肚即是已炸透的魚肚，耐存，用薑、葱、酒及白醋煨煮片刻，可去掉腥味。

Fried fish maw is fish maw that has been deep-fried. It has a long duration. Cook the fish maw with ginger, spring onion, wine and white vinegar for a moment. Its fishy smell can be removed.

柱候牛腩煲
Beef Brisket with Chu Hou Sauce

材料 Ingredients

牛坑腩 2 斤
蘿蔔 2 斤
柱候醬 6 湯匙
薑 8 片
青蒜 3 棵
蒜茸 1 湯匙
水 10 杯

1.2 kg beef brisket point-end
1.2 kg white radish
6 tbsps Chu Hou sauce
8 slices ginger
3 stalks green garlic
1 tbsp finely chopped garlic
10 cups water

調味料 Seasoning

冰糖 1 1/2 湯匙
桔餅 1 個
陳皮半個

1 1/2 tbsps rock sugar
1 sugar-preserved tangerine
1/2 dried tangerine peel

獻汁 Thickening glaze

生粉 1 茶匙
水 1 湯匙
* 拌勻

1 tsp caltrop starch
1 tbsp water
* mixed well

做法 Method

1. 牛坑腩洗淨，放入水及白醋 1 湯匙的鍋內，煲約 15 分鐘，取出，沖淨，切件備用。

2. 陳皮用水浸軟，刮去瓤、切絲；青蒜切成短度；蘿蔔去皮、切件。

3. 燒熱鑊，下少許油，放入青蒜、薑片、蒜茸、柱候醬及牛腩爆香，潷酒，加入調味料及水煮滾，轉放瓦煲內，以慢火燜約 1 小時，關火，焗約 4 至 5 小時，再放入蘿蔔件燜 30 分鐘，灑入鹽再燜片刻，潷酒，埋獻即可。

1. Rinse the beef brisket. Put in a pot of water with 1 tbsp of white vinegar. Cook for about 15 minutes. Take out and rinse. Cut into pieces. Set aside.

2. Soak the dried tangerine peel in water until soft. Scrape off the pith. Cut into shreds. Cut the green garlic into short sections. Peel the white radish and cut into pieces.

3. Heat a wok. Add a little oil. Put in the green garlic, ginger, garlic, Chu Hou sauce and beef brisket. Stir-fry until fragrant. Sprinkle with wine. Add the seasoning and water. Bring to the boil. Transfer to a casserole. Simmer for about 1 hour. Turn off heat. Leave for about 4 to 5 hours with the lid on. Put in the white radish and simmer for 30 minutes. Season with salt and simmer for a while. Sprinkle with wine. Mix in the thickening glaze. Serve.

砂窩雲吞雞

Wanton and Chicken in Casserole

材料 *Ingredients*

雞 1 隻（約 2 斤）
金華火腿 2 兩
瑤柱 3 粒
菜膽 4 兩
薑 3 片

1 chicken (about 1.2 kg)
75 g Jinhua ham
3 dried scallops
150 g vegetables (tender part)
3 slices ginger

雲吞材料 *Ingredients for wanton*

蝦 1 斤
豬肉半斤
雲吞皮 6 兩

600 g shrimps
300 g pork
225 g wanton wrappers

雲吞調味料 *Seasoning for wanton*

鹽 1 1/4 茶匙
生粉 1 湯匙
大地魚茸 2 茶匙
炒香芝麻 1 茶匙
蛋黃 1 個
糖、胡椒粉及麻油各少許

1 1/4 tsps salt
1 tbsp caltrop starch
2 tsps finely chopped dried plaice
1 tsp toasted sesame seeds
1 egg yolk
sugar
ground white pepper
sesame oil

小秘訣 Tips

當雲吞煮至浮起時，加入清水半碗同煮，令雲吞爽口，而且不黏鍋。

When the wantons float on boiling water, add 1/2 bowl of water to cook with. The wantons will be crunchy and not stick to the pot.

雲吞做法
Method for wanton

1. 蝦去殼、挑腸，用生粉及鹽洗淨，抹乾水分，備用。
2. 豬肉洗淨，抹乾，剁粒，加入蝦及調味料順一方向拌勻，撻至起膠帶黏性，放入雪櫃待用。
3. 用雲吞皮包入餡料，黏緊，放入滾水內煮至浮起，加入水半碗再煮滾，瀝乾水分備用。

1. Shell and devein the shrimps. Clean with caltrop starch and salt. Rinse and wipe dry. Set aside.
2. Rinse and wipe dry the pork. Chop into dices. Add the shrimps and seasoning. Stir evenly in one direction. Throw into a bowl repeatedly until sticky. Chill in a refrigerator.
3. Wrap the stuffing in wanton wrappers. Seal by pressing the edges tightly. Put into boiling water and cook until they float. Add 1/2 bowl of water. Bring to the boil. Drain. Set aside.

做法
Method

1. 雞洗淨，燒熱一鍋滾水，放入雞略飛水，用水沖淨。
2. 金華火腿飛水，洗淨，切絲；瑤柱用水浸軟，瑤柱水留用。
3. 砂鍋內放入雞、清水 16 杯、瑤柱（連瑤柱水）、金華火腿及薑片，煲約 2 小時，下菜膽及煮熟雲吞，最後下鹽、紹酒及麻油略拌即可。

1. Wash the chicken. Bring a pot of water to the boil. Blanch the chicken. Rinse well.
2. Blanch the Jinhua ham and rinse. Cut into shreds. Soak the dried scallops in water until soft. Keep the dried scallop water.
3. Put the chicken, 16 cups of water, dried scallops (with dried scallop water), Jinhua ham and ginger into a casserole. Cook for about 2 hours. Add the vegetables and cooked wantons. Finally add the salt, Shaoxing wine and sesame oil. Stir slightly. Serve.

梅菜扣肉

Stewed Pork Belly with
Preserved Flowering Cabbage

材料
Ingredients

五花腩肉 1 斤
甜梅菜半斤
白米醋 1 湯匙

600 g pork belly
300 g sweetened preserved
flowering cabbage
1 tbsp white rice vinegar

調味料
Seasoning

老抽及生抽各 2 湯匙
鹽半茶匙
片糖碎 1 塊
紹酒 1 湯匙
麻油 1 茶匙

2 tbsps dark soy sauce
2 tbsps light soy sauce
1/2 tsp salt
1 slab sugar (crushed)
1 tbsp Shaoxing wine
1 tsp sesame oil

做法
Method

1. 燒滾水，下白米醋 1 湯匙及腩肉煲約 15 分鐘，用水沖片刻，抹乾水分。用老抽 1 湯匙塗抹腩肉，在皮部用叉戳孔，炸至金黃色，用水略沖，抹乾後切件。

2. 梅菜用水略浸，擠乾水分，切絲，用油及糖各 1 茶匙炒乾及散出香味，放入鉢頭內鋪好，排上腩肉（皮向上），傾入調味料，隔水蒸約 2 1/2 小時至軟，即可享用。

1. Bring water to the boil. Put in 1 tbsp of white rice vinegar and the pork belly. Boil for about 15 minutes. Rinse for a while. Wipe dry. Spread 1 tbsp of dark soy sauce on the pork belly. Pierce the skin with a fork. Deep-fry until golden. Slightly rinse. Wipe dry and cut into pieces.

2. Soak the preserved flowering cabbage in water for a while. Squeeze water out. Cut into shreds. Stir-fry with 1 tsp of oil and 1 tsp of sugar until dry and sweet-scented. Lay in an earthen bowl. Arrange the pork belly on top (skin up). Pour in the seasoning. Steam for about 2 1/2 hours, or until tender. Serve.

小秘訣 Tips

- 五花腩肉外皮較厚，用叉略戳孔有效將調味料滲入。
- 梅菜必須用水略浸，再用油及糖炒乾，可散發梅菜獨有之香味。
- The skin of pork belly is quite thick. Pierce with a fork will help the seasoning permeate through the holes.
- Preserved flowering cabbage must be soaked in water for a while, and then stir-fried with oil and sugar until dry to let its unique aroma diffuse.

檸檬芥汁雞
Fried Chicken in Lemon Mustard Sauce

材料 Ingredients

雞胸肉 6 件
蒜肉 2 粒
乾葱肉 2 粒
上湯 3/4 杯

pcs chicken breast
cloves skinned garlic
shallots
/4 cup stock

醃料 Marinade

末粉 2 茶匙
茸粉 1 茶匙
半茶匙
檬汁及糖各 1 1/2 茶匙
椒粉少許

tsps mustard powder
tsp garlic powder
2 tsp salt
1/2 tsps lemon juice
1/2 tsps sugar
ound white pepper

調味料 Seasoning

末粉 1 茶匙
2 茶匙
檬汁 1/4 杯
半茶匙
椒粉少許

tsp mustard powder
tsps sugar
4 cup lemon juice
2 tsp salt
ound white pepper

做法 Method

1. 雞胸肉去皮，用刀背略剁，加入醃料拌勻，待 1 小時，放入平底鑊內煎至兩面金黃色，潷酒上碟。

2. 燒熱油 1 湯匙，爆香乾葱肉及蒜肉，傾入上湯，下調味料煮勻，取出乾葱及蒜肉，用生粉水埋獻，將醬汁澆在雞肉上即可。

1. Skin the chicken breast. Slightly chop with the back of a knife. Mix with the marinade and rest for 1 hour. Fry on a pan until both sides are golden. Sprinkle with wine. Put on a plate.

2. Heat 1 tbsp of oil. Stir-fry the shallots and garlic until aromatic. Pour in the stock. Add the seasoning and bring to the boil. Take out the shallots and garlic. Thicken the sauce with caltrop starch solution. Pour the sauce on the chicken. Serve.

小秘訣 Tips

不喜歡雞胸肉的質感，建議選用雞髀肉或雞翼皆可，雞髀肉滑嫩多肉。

If you dislike the texture of chicken breast, you may use boned chicken leg or chicken wing. Boned chicken leg is smooth and meaty.

燜鴨舌
Stewed Duck Tongues

材料
Ingredients

鴨舌 450 克
薑 2 片
葱 2 條（切短度）
蒜肉 1 粒

450 g duck tongues
2 slices ginger
2 sprigs spring onion (cut into short sections)
1 clove skinned garlic

調味料
Seasoning

蠔油 1 湯匙
生抽 2 湯匙
老抽半湯匙
糖 2 茶匙
麻油 1/4 茶匙
八角 6 小角
桂皮 1 小塊
水 5 安士（約 145 毫升）

1 tbsp oyster sauce
2 tbsps light soy sauce
1/2 tbsp dark soy sauce
2 tsps sugar
1/4 tsp sesame oil
6 star anise segments
1 small piece cinnamon
5 oz. (145 ml) water

做法
Method

1. 鴨舌解凍，洗淨，放入滾水略灼，去衣，用鹽 1 茶匙擦洗，沖水，瀝乾水分。

2. 燒熱油 2 湯匙，放入薑片、葱段及蒜肉爆香，下鴨舌略拌，潷酒，隨即放入調味料煮滾，用慢火燜 10 分鐘，最後下少許麻油，冷熱吃皆可。

1. Defrost the duck tongues. Rinse well. Slightly blanch and remove the skin. Rub with 1 tsp salt. Rinse and drain.

2. Heat 2 tbsps of oil. Stir-fry the ginger, spring onion and garlic until fragrant. Add the duck tongues and roughly stir-fry. Sprinkle with wine. Pour in the seasoning and bring to the boil. Simmer for 10 minutes. Finally sprinkle with a little sesame oil. Serve hot or cold.

小秘訣 Tips

每粒八角呈八小角，只取用六小角，香氣已非常足夠！

Each star anise has eight segments. Only six segments can make the dish smell brilliant!

乾 煎 明 蝦
Fried Prawns with Ketchup

材料 Ingredients

大蝦 1 斤
蒜肉 2 粒（剁茸）

600 g prawns
2 cloves skinned garlic (finely chopped)

調味料 Seasoning

糖 2 茶匙
生抽半茶匙
鹽 1/4 茶匙
茄汁 2 湯匙
胡椒粉及麻油各少許

2 tsps sugar
1/2 tsp light soy sauce
1/4 tsp salt
2 tbsps ketchup
ground white pepper
sesame oil

小秘訣 Tips

蝦肉容易熟透，見蝦變成紅色及呈彎曲狀，再略煎即可。

It is easy for the prawns to be cooked through. When they turn red and curl up, fry a little longer. It is done.

做法 Method

1. 大蝦的鬚腳用剪刀去掉，挑腸，用鹽水洗淨，備用。
2. 燒熱鑊，下油 4 湯匙，放入大蝦煎至兩面熟透，盛起。
3. 燒熱油，下蒜茸爆香，放入大蝦回鑊炒勻，潷酒，傾入調味料拌勻上碟。

1. Cut away the legs and tentacles of the prawns with scissors. Devein. Clean with salted water. Set aside.
2. Heat a wok. Put in 4 tbsps of oil. Fry both sides of the prawns until fully cooked. Set aside.
3. Heat oil. Stir-fry the garlic until aromatic. Return the prawns and stir-fry evenly. Sprinkle with wine. Pour in the seasoning and mix evenly. Serve.

京香骨
Stewed Pork Spareribs in Vinegar Sauce

材料 Ingredients

腩排 12 兩
薑 4 片

450 g pork spareribs
4 slices ginger

調味料 Seasoning

鎮江醋 4 湯匙
鹽 1 茶匙
糖 4 湯匙
水適量
老抽及茄汁各 1 湯匙（後下）
麻油少許（後下）

4 tbsps Zhenjiang vinegar
1 tsp salt
4 tbsps sugar
water
1 tbsp dark soy sauce (added last)
1 tbsp ketchup (added last)
sesame oil (added last)

做法 Method

1. 腩排斬成約 2 吋長段，放入熱油內略泡油，瀝乾油分。
2. 燒熱少許油，下薑片爆香，放入腩排炒勻，加入調味料蓋過腩排表面，燜約 40 分鐘至水分收乾。
3. 上碟前，加入老抽、茄汁及麻油拌勻即可。

1. Chop the spareribs into sections of about 2 inches in length. Deep-fry in hot oil for a while. Drain.
2. Heat a little oil. Stir-fry the ginger until fragrant. Put in the spareribs and stir-fry evenly. Add the seasoning. The liquid should be enough to cover the spareribs. Stew for about 40 minutes, or until the sauce dries.
3. Mix in the dark soy sauce, ketchup and sesame oil right before serving.

小秘訣 Tips

由於下鎮江醋燜煮，建議用瓦鍋烹調，若用金屬器具煮帶酸性的醋，或會產生化學變化。

Cooking this dish in a casserole is recommended as Zhenjiang vinegar is used. The acidity of vinegar may cause chemical change if cooked in metal cookware.

威化香蕉蝦

Banana and Prawn Wrapped in Wafer Paper

小秘訣 Tips

威化紙容易受潮，宜置於乾爽地方，於售賣粉麵的店鋪或烘焙食品店有售。

It is easy for the wafer paper to get moist. Place it in a dry, ventilated area. It can be bought at shops selling noodles or baking ingredients.

材料 Ingredients

中蝦 10 隻
香蕉 2 隻
威化紙數張
沙律醬 4 湯匙

10 medium prawns
2 bananas
several sheets wafer paper
4 tbsps salad dressing

調味料 Seasoning

鹽 1/4 茶匙
蛋白 1 湯匙
生粉、胡椒粉及麻油
各少許

1/4 tsp salt
1 tbsp egg white
caltrop starch
ground white pepper
sesame oil

做法 Method

1. 蝦洗淨，去頭、去殼，保留尾部，挑腸，切成雙飛狀，拌入調味料。
2. 香蕉去皮，切成薄塊；威化紙剪裁成長方塊。
3. 威化紙鋪平，包入蝦及香蕉（蝦尾在外），捲起，以生粉水塗抹邊沿，緊黏封口。
4. 燒熱油，放入威化蝦卷炸至金黃及熟透，瀝乾油分，蘸沙律醬享用。

1. Wash the prawns. Remove the head and shell. Keep the tail. Devein and cut along the back of the prawns. Mix in the seasoning.
2. Skin the bananas. Finely slice. Trim the wafer paper into rectangles.
3. Lay the wafer paper flat. Wrap the banana and prawn (tail outside). Roll up. Spread caltrop starch solution along the edge. Stick firmly to seal the opening.
4. Heat oil. Deep-fry the prawn rolls until golden and fully cooked. Drain. Serve with the salad dressing.

紗紙鹽焗雞

Paper-wrapped Salt-roasted Chicken

材料 Ingredients

雞 1 隻（2 斤）
薑 4 片
葱 2 條
紗紙 2 張

1 chicken (about 1.2 kg)
4 slices ginger
2 sprigs spring onion
2 sheets mulberry paper

調味料 Seasoning

沙薑粉 3 湯匙
鹽 2 湯匙
麻油 1 茶匙
紹酒 1 茶匙

3 tbsps sand ginger powder
2 tbsps salt
1 tsp sesame oil
1 tsp Shaoxing wine

小秘訣 Tips

另可將大量粗鹽炒至黃色，包入雞焗熟或用焗爐烤熟，但以蒸的方法最能保持雞肉滑嫩。

You may choose to stir-fry a large amount of coarse salt until brown, and then put the chicken inside to roast until done. Or bake the chicken in an oven. But the best way to keep the meat texture smooth is by steaming.

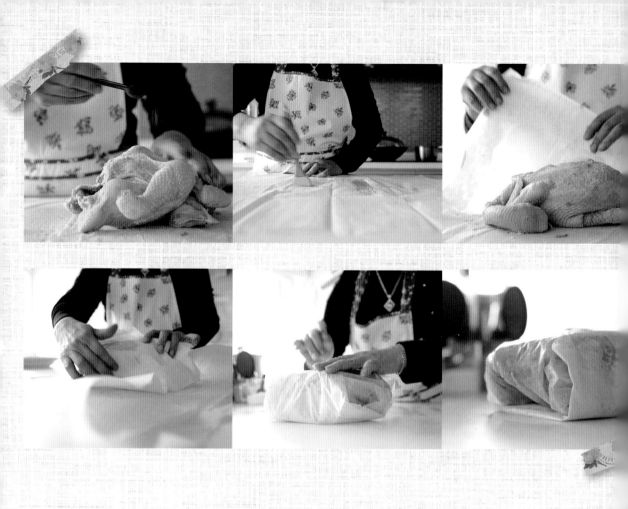

做法
Method

1. 雞洗淨，抹乾水分備用。

2. 於雞腔內外均勻地抹上調味料，並放入薑蔥於雞腔內，醃半小時。

3. 紗紙掃上油，放上雞包好，隔水大火蒸 25 分鐘，待涼，斬件上碟。

1. Wash the chicken. Wipe dry and set aside.

2. Spread the seasoning evenly on the chicken inside and outside. Put the ginger and spring oni
   inside the chicken. Marinate for 1/2 hour.

3. Brush oil on the mulberry paper. Put the chicken on the paper. Wrap up. Steam over high h
   for 25 minutes. Let it cool down. Chop into pieces. Serve.

燴莧菜蛋角
Fried Chinese Spinach Egg Dumplings in Soup

材料 Ingredients

莧菜半斤
蝦膠 6 兩（做法參考 P.14）
雞蛋 5 個
金華火腿茸 3 湯匙

300 g Chinese spinach
225 g minced shrimp
(refer to p.14 for method)
5 eggs
3 tbsps finely chopped Jinhua ham

蝦膠調味料 Seasoning for minced shrimp

鹽 1/3 茶匙
蛋白 1 湯匙
胡椒粉及麻油各少許

1/3 tsp salt
1 tbsp egg white
ground white pepper
sesame oil

調味料 Seasoning

鹽半茶匙
糖 1/4 茶匙
胡椒粉少許

1/2 tsp salt
1/4 tsp sugar
ground white pepper

上湯料 Ingredients for soup

上湯 3 杯
蠔油 2 湯匙
胡椒粉及麻油各少許

3 cups stock
2 tbsps oyster sauce
ground white pepper
sesame oil

小秘訣 Tips

金華火腿鹹味太濃的話，可放入滾水煮 15 分鐘，洗淨，加少許糖及紹酒蒸 15 分鐘，去除鹹味及油羶味。

If the Jinhua ham is too salty, put it in boiling water for 15 minutes and then rinse it. After that, steam the Jinhua ham with a little sugar and Shaoxing wine for 15 minutes. It can help remove its saltiness and oily smell.

1. 莧菜去根部及葉，洗淨，摘取幼嫩部分，放入滾水內，下鹽及油飛水，瀝乾水分，切碎備用。

2. 蝦膠與調味料順一方向拌勻，備用。

3. 雞蛋拂勻，下調味料，拌入蝦膠、莧菜茸及金華火腿茸。

4. 燒熱鑊，下油3湯匙，舀入一勺蝦膠莧菜蛋液，煎至兩面金黃色，瀝乾油分。

5. 煮滾上湯料，放入全部蛋角燴煮，上碟即成。

1. Remove the roots and leaves of the Chinese spinach. Rinse and pick the tender part. Blanch in boiling water with salt and oil. Drain and chop up. Set aside.

2. Stir the minced shrimp with seasoning evenly in one direction. Set aside.

3. Whisk the eggs. Add the seasoning. Fold in the minced shrimp, Chinese spinach and Jinhua ham. Mix well.

4. Heat a wok. Put in 3 tbsps of oil. Pour in one ladle of the egg mixture. Fry until both sides are golden. Drain.

5. Bring the soup ingredients to the boil. Put in the fried egg dumplings. Cook for a while. Serve.

椒鹽白飯魚

Deep-fried Chinese Noodlefish with Chilli and Spice Salt

材料 Ingredients

白飯魚 10 兩
麵粉 6 湯匙
紅辣椒 1 隻（切粒）
淮鹽少許

375 g Chinese noodlefish
6 tbsps flour
1 red chilli (diced)
spice salt

醃料 Marinade

鹽及胡椒粉各半茶匙

1/2 tsp salt
1/2 tsp ground white pepper

做法 Method

1. 白飯魚洗淨，抹乾水分，用醃料略醃。
2. 白飯魚與麵粉拌勻，放入熱油內炸至金黃香脆，盛起，灑上淮鹽及紅椒碎即可享用。

1. Wash the fish. Wipe dry. Roughly mix with the marinade.
2. Mix the fish with flour evenly. Deep-fry in hot oil until golden and crisp. Dish up. Sprinkle with the spice salt and red chilli. Serve.

小秘訣 Tips

- 白飯魚與麵粉拌勻才炸，脆卜香口，用粟粉絕對不能達到此效果。
- 白飯魚必須抹乾才拌麵粉，否則黏黏的糊成一團。
- Chinese noodlefish will be crisp and delicious by mixing with plain flour before deep-frying. Corn flour can never do that.
- The fish must be dried before combining with flour; otherwise, it will be sticky and mushy.

蠔豉蒸肉餅
Steamed Minced Pork with Dried Oysters

材料
Ingredients

半肥瘦豬肉半斤
蠔豉 2 兩（用水浸軟）
馬蹄粒 4 湯匙
葱白粒 1 湯匙
薑 4 片
葱 3 條

300 g half lean and half fat pork
75 g dried oysters
 (soaked in water to soften)
4 tbsps diced water chestnuts
1 tbsp diced spring onion (white part)
4 slices ginger
3 sprigs spring onion

醃料
Marinade

油、鹽及生抽各 1 茶匙
薑汁 2 茶匙
糖半茶匙
胡椒粉、麻油及紹酒各少許

1 tsp oil
1 tsp salt
1 tsp light soy sauce
2 tsps ginger juice
1/2 tsp sugar
ground white pepper
sesame oil
Shaoxing wine

生粉水
Caltrop starch solution

生粉 1 湯匙
水 3 湯匙
* 拌勻

1 tbsp caltrop starch
3 tbsps water
* mixed well

做法
Method

1. 燒熱鑊，下油半茶匙，爆香薑片及葱，灒酒，傾入水及蠔豉煨約 5 分鐘，取出洗淨，備用。
2. 豬肉洗淨，抹乾水分，切粒，再剁成肉餅狀，加入蠔豉略剁，拌入馬蹄粒及葱白。
3. 加入醃料順一方向拌勻，最後下全部生粉水攪至起膠，隔水蒸約 8 至 10 分鐘即成。

1. Heat a wok. Put in 1/2 tsp of oil. Stir-fry the ginger and spring onion until sweet-scented. Sprinkle with wine. Pour in water and the dried oysters. Cook for 5 minutes. Take out and rinse. Set aside.
2. Rinse the pork. Wipe dry. Dice and then chop into minced pork. Add the dried oysters and chop roughly. Mix in the water chestnuts and white part of spring onion.
3. Add the marinade. Stir in one direction. Finally mix in the caltrop starch solution until gluey. Steam for about 8 to 10 minutes. Serve.

小秘訣 Tips

蠔豉先用薑、葱及酒煨 5 分鐘，可去除蠔腥味。

Cook the dried oysters with ginger, spring onion and wine for 5 minutes to remove their fishy smell.

鹽燒三文魚頭
Salt-roasted Salmon Head

材料 Ingredients

三文魚頭（大）1 個
蛋白 1 個

1 large salmon head
1 egg white

調味料 Seasoning

鹽 2 茶匙
胡椒粉適量

2 tsps salt
ground white pepper

做法 Method

1. 三文魚頭洗淨，抹乾水分，斬成兩件。
2. 蛋白拂勻，塗抹在魚頭上，灑上鹽及胡椒粉調味。
3. 三文魚頭鋪在錫紙上，放入焗爐以 220℃ 烤焗，見鹽粒漸漸變成白色及魚肉熟透即成。

1. Wash the salmon head. Wipe dry. Chop into two pieces.
2. Whisk the egg white. Spread on the salmon head. Season with salt and ground white pepper.
3. Put the salmon head on aluminum foil. Put in an oven and bake at 220°C. When the salt gradually turns white and the fish meat is cooked through. Serve.

小秘訣 Tips

別擔心鹹味過重，因鹽粒烤至轉成白色，會自然脫落。

Don't worry about how salty it is. When the salt turns white in the baking process, it will fall off.

蜜汁雞翼
Honey-glazed Chicken Wings

小秘訣 Tips

將急凍雞翼浸於淡鹽水內，可加速解凍及去除雪味。

Soaking the frozen chicken wings into light salted water will expedite the defrosting process and remove the unpleasant smell.

輕鬆上桌

材料 Ingredients

雞翼 1 斤
蔥 2 條（切段）

600 g chicken wings
2 sprigs spring onion
(cut into sections)

醃料 Marinade

鹽半茶匙
生抽 2 茶匙
紹酒、胡椒粉及麻油各少許

1/2 tsp salt
2 tsps light soy sauce
Shaoxing wine
ground white pepper
sesame oil

調味料 Seasoning

生抽 1 茶匙
糖 1 1/2 茶匙
蜜糖 2 湯匙
暖水 2 湯匙
* 拌勻

1 tsp light soy sauce
1 1/2 tsps sugar
2 tbsps honey
2 tbsps warm water
* mixed well

做法 Method

1. 雞翼用少許鹽略擦，沖洗，瀝乾水分，下醃料醃透。
2. 燒熱油，下雞翼炸至微黃色，瀝乾油分。
3. 燒熱少許油，下蔥段爆香，取走蔥段，放入調味料煮滾，下雞翼略煮至吸收汁液，上碟即可。

1. Clean the chicken wings with a little salt. Rinse and drain. Mix with the marinade and rest for a longer time.
2. Heat oil. Deep-fry the chicken wings until light brown. Drain.
3. Heat a little oil. Stir-fry the spring onion until aromatic. Remove the spring onion. Put in the seasoning and bring to the boil. Add the chicken wings and cook for a while. When the chicken wings absorb the sauce, serve.

鮑汁素燒鵝
Vegetarian Roast Goose in Abalone Sauce

材料 Ingredients

腐皮 4 張

4 sheets beancurd skin

上湯 Stock

鮑汁 2 湯匙
生抽 1 茶匙
老抽 1 茶匙
麻油 1 茶匙
水 1 杯

2 tbsps abalone sauce
1 tsp light soy sauce
1 tsp dark soy sauce
1 tsp sesame oil
1 cup water

浸汁料 Marinating sauce

鮑汁 4 湯匙
生抽、老抽及麻油各半茶匙
水半杯
* 拌勻

4 tbsps abalone sauce
1/2 tsp light soy sauce
1/2 tsp dark soy sauce
1/2 tsp sesame oil
1/2 cup water
* mixed well

小秘訣 Tips

煎炸的素燒鵝香脆可口；蒸吃的則較清淡，悉隨尊便。

Deep-fried vegetarian roast goose is crunchy and yummy whereas the steamed is light in taste. Choose the one you like.

做法
Method

1. 腐皮用濕布抹掃，剪掉硬邊，均勻地掃上浸汁料。將兩張腐皮重疊，摺成長方形，用牙籤緊緊串起，其餘兩張腐皮依法處理。

2. 將已捲好的素燒鵝浸於浸汁料內，待約 30 分鐘，取出，蒸 10 分鐘，瀝乾水分，放於乾爽地方徹底乾透。

3. 燒熱油，放入已乾透的素燒鵝，煎炸至表面香脆，切件，排於碟上。

4. 煮滾上湯，澆於素燒鵝上即可。

1. Clean the beancurd skin with a damp cloth. Cut away the tough edge. Brush the marinatin sauce on the beancurd skin evenly. Put 2 sheets of beancurd skin together, one over anothe Fold into a rectangular shape. Skewer tightly with toothpicks. Do the same to the rest 2 shee of beancurd skin.

2. Soak the folded beancurd skin in the marinating sauce. Rest for about 30 minutes. Take out an steam for 10 minutes. Drain well. Place in a dry, ventilated area to let it dry entirely.

3. Heat oil. Deep-fry the fully dried beancurd skin until the surface is crunchy. Cut into piece Arrange on a plate.

4. Bring the stock to the boil. Pour over the vegetarian roast goose. Serve.

家鄉釀涼瓜
Stuffed Bitter Cucumber

材料
Ingredients

鯪魚膠半斤
梅頭肉 4 兩（攪碎）
鹹蛋黃 1 個（切粒）
蝦醬 1 1/2 湯匙
涼瓜（長型）1 斤
豆豉 1 1/2 湯匙
蒜茸 1 湯匙
葱粒 2 湯匙

300 g mud carp paste
150 g pork collar-butt (minced)
1 salted egg yolk (diced)
1 1/2 tbsps shrimp paste
600 g bitter cucumber (long shape)
1 1/2 tbsps fermented black beans
1 tbsp finely chopped garlic
2 tbsps finely diced spring onion

餡肉調味料
Seasoning for stuffing

鹽 1 1/2 茶匙
生粉 1 湯匙
胡椒粉少許
水 3 湯匙
葱粒 2 湯匙

1 1/2 tsps salt
1 tbsp caltrop starch
ground white pepper
3 tbsps water
2 tbsps diced spring onion

調味料
Seasoning

鹽半茶匙
片糖半塊
水 2 1/2 杯

1/2 tsp salt
1/2 slab sugar
2 1/2 cups water

用片糖燜煮涼瓜，可去除涼瓜的苦澀味道。

To stew the bitter cucumber with slab sugar can help remove its bitter taste.

做法
Method

1. 鯪魚膠及梅頭肉順一方向拌勻，加入餡肉調味料撻至起膠及有黏性。

2. 涼瓜洗淨，切成 1 1/2 吋厚圓形，去核，放入滾水煮約 5 分鐘，過冷河，瀝乾水分。

3. 蝦醬、紹酒少許及生粉 2 茶匙拌勻，在涼瓜環內部抹上少許蝦醬，釀入餡料，鋪上鹹蛋黃。

4. 燒熱鑊，下油 3 湯匙，放入涼瓜用慢火煎至兩面金黃色。

5. 燒熱鑊，下少許油爆香蒜茸及豆豉，放入涼瓜環及調味料略煮，潷酒，加蓋慢火燜至熟透，即可上碟。

1. Stir the mud carp paste and pork collar-butt evenly in one direction. Add the seasoning for stuffing. Throw into a bowl repeatedly until sticky.

2. Rinse the bitter cucumber. Cut into round pieces of 1 1/2 inches thick. Remove the seeds. Cook in boiling water for about 5 minutes. Rinse in cold water. Drain.

3. Combine the shrimp paste, a little Shaoxing wine and 2 tsps of caltrop starch together. Spread a little shrimp paste inside the bitter cucumber. Fill in the stuffing. Lay the salted egg yolk on top.

4. Heat a wok. Put in 3 tbsps of oil. Fry the bitter cucumber over low heat until both sides are golden.

5. Heat a wok. Put in a little oil. Stir-fry the garlic and fermented black beans until aromatic. Put in the stuffed bitter cucumber and seasoning. Cook for a while. Sprinkle with wine. Put a lid on and simmer until done. Serve.

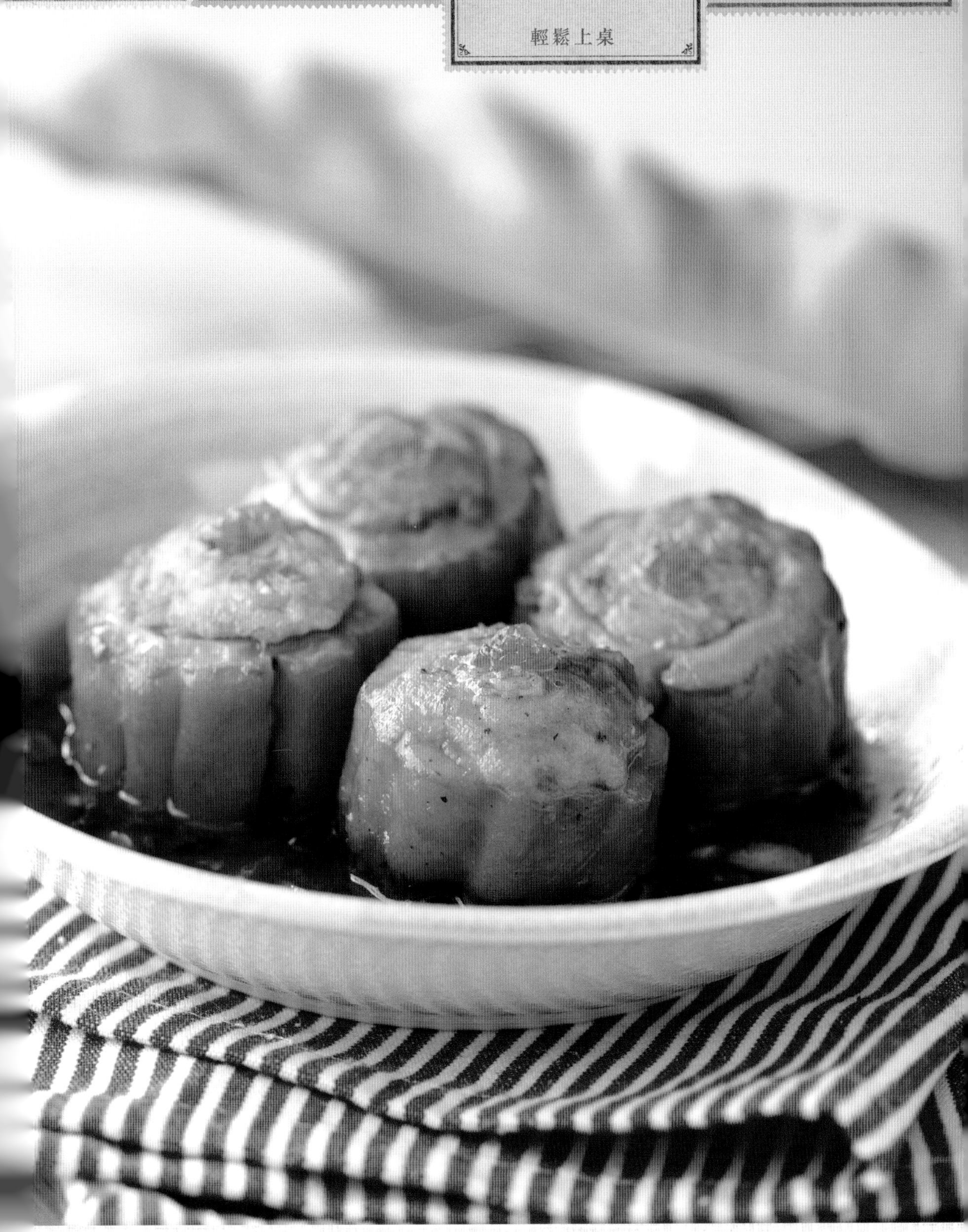

黑椒瀬尿蝦
Stir-fried Mantis Shrimps with Black Peppercorns

輕鬆上桌

材料 Ingredients

瀨尿蝦 12 兩
紅辣椒 1 隻（切絲）
蒜茸 1 茶匙

450 g mantis shrimps
1 red chilli (shredded)
1 tsp finely chopped garlic

調味料 Seasoning

黑椒粒及鹽各半茶匙
胡椒粉、麻油及水各少許

1/2 tsp black peppercorns
1/2 tsp salt
ground white pepper
sesame oil
water

做法 Method

1. 瀨尿蝦盛於笓箕內，加入粗鹽 1 湯匙輕輕搖動，放於水喉下沖淨，瀝乾水分。
2. 瀨尿蝦撲上生粉，放入溫油內泡油，瀝乾油分。
3. 燒熱油 2 湯匙，下蒜茸及紅椒絲爆香，放入瀨尿蝦回鑊，潷酒，下調味料大火炒至乾身，即可上碟。

1. Put the mantis shrimps in a strainer. Add 1 tbsp of coarse salt. Gently shake. Rinse under running tap water. Drain.
2. Coat the mantis shrimps with caltrop starch. Deep-fry in warm oil for a moment. Drain.
3. Heat 2 tbsps of oil. Stir-fry the garlic and red chilli until aromatic. Return the mantis shrimps. Sprinkle with wine. Add the seasoning. Stir-fry over high heat until the mantis shrimp dries. Put on a plate. Serve.

小秘訣 Tips

瀨尿蝦與粗鹽拌勻後輕輕搖動，令蝦隻分泌尿液。

Mix the mantis shrimps with coarse salt and then give it a gently shake. The mantis shrimps will urinate.

欖角頭抽燜腩肉

材料 Ingredients

腩肉 1 斤
欖角 4 湯匙
薑 3 片
陳皮半個

600 g pork belly
4 tbsps preserved black olives
3 slices ginger
1/2 dried tangerine peel

調味料 Seasoning

鹽半茶匙
頭抽 4 湯匙
老抽、麻油及冰糖各 1 湯匙
水 2 杯

1/2 tsp salt
4 tbsps premium light soy sauce
1 tbsp dark soy sauce
1 tbsp sesame oil
1 tbsp rock sugar
2 cups water

做法 Method

1. 陳皮用水浸軟、去瓤,切絲備用。
2. 腩肉洗淨,放入煲內,加清水及白醋 1 湯匙煲約 5 分鐘,過冷河備用。
3. 調味料放入煲內,下薑片、陳皮絲及欖角 2 湯匙,煲約 5 分鐘,放入原件腩肉煲 20 分鐘,取出待涼,切件,放回煲內,並加入餘下欖角及麻油 1 茶匙,再煲 5 分鐘即成。

1. Soak the dried tangerine peel in water to soften. Remove the pith. Cut into shreds. Set aside.
2. Wash the pork belly. Put into a pot. Add water and 1 tbsp of white vinegar. Cook for about 5 minutes. Rinse in cold water. Set aside.
3. Put the seasoning in the pot. Add the ginger, dried tangerine peel and 2 tbsps of preserved black olives. Cook for about 5 minutes. Put in the whole pork belly and cook for 20 minutes. Take out. Let it cool down. Cut into pieces. Return to the pot. Add the rest black olives and 1 tsp of sesame oil. Cook for 5 minutes. Serve.

小秘訣 Tips

• 腩肉原件燜煮,緊鎖肉汁,也避免肉質收縮。
• 最後灑上欖角略煮,可保留欖角的香味。
• Braising the whole pork belly can lock the juice in and avoid shrinking.
• Cooking some black olives at the final stage can keep their aroma.

Braised Pork Belly with Preserved Olives
in Premium Light Soy Sauce

咖喱抄手
Curry Wantons

材料 Ingredients

蝦肉 6 兩
梅頭肉 4 兩
筍肉 4 兩
水餃皮 12 兩
咖喱粉 1 湯匙
乾葱 2 粒（切粒）
葱粒 4 湯匙

225 g shelled shrimps
150 g pork collar-butt
150 g skinned bamboo shoot
450 g dumpling wrappers
1 tbsp curry powder
2 shallots (diced)
4 tbsps diced spring onion

調味料 Seasoning

花奶 2 湯匙
上湯半杯
糖 1/4 茶匙
粟粉 1 茶匙
鹽、生抽及麻油各少許

2 tbsps evaporated milk
1/2 cup stock
1/4 tsp sugar
1 tsp corn flour
salt
light soy sauce
sesame oil

醃料 Marinade

生抽及鹽各 1 茶匙
粟粉 2 湯匙
糖、麻油及胡椒粉各少許

1 tsp light soy sauce
1 tsp salt
2 tbsps corn flour
sugar
sesame oil
ground white pepper

做法 Method

1. 蝦肉用鹽及生粉洗淨，抹乾水分；梅頭肉洗淨，抹乾，用刀剁碎；筍肉飛水，切幼條。
2. 將蝦肉、肉碎、筍肉及醃料拌勻，撻至起膠及帶黏性。
3. 用水餃皮包入餡料，按緊皮邊，放入滾水內煮至浮起，加入水半碗再煮滾，盛起，瀝乾水分，上碟備用。
4. 燒熱少許油，下乾葱及咖喱粉爆香，潷酒，加入調味料煮勻至略稠，澆在雲吞面，灑上葱粒即可。

1. Clean the shelled shrimps with salt and caltrop starch. Rinse and wipe dry. Rinse the pork collar-butt. Wipe dry and finely chop. Blanch the bamboo shoot. Cut into thin strips.
2. Mix the shrimps, pork, bamboo shoot and marinade evenly. Throw into a bowl repeatedly until sticky.
3. Wrap the stuffing in the dumpling wrapper. Press the edges firmly. Cook in boiling water until they float. Add 1/2 bowl of water. Bring to the boil. Drain. Put on a plate.
4. Heat a little oil. Stir-fry the shallots and curry powder until fragrant. Sprinkle with wine. Put in the seasoning and cook until the sauce is a bit thick. Pour on the wantons. Sprinkle with the spring onion. Serve.

古法蒸雞
Traditional Steamed Chicken

材料 Ingredients

雞半隻
冬菇5個（處理方法參考 p.12）
金華火腿 2 湯匙
紅棗 8 粒（去核）
大頭菜 1 片
薑 6 片
葱 1 條

1/2 chicken
5 dried black mushrooms
(refer to p.12 for method)
2 tbsps Jinhua ham
8 red dates (cores removed)
1 slice preserved mustard root
6 slices ginger
1 sprig spring onion

調味料 Seasoning

鹽半茶匙
生抽 2 茶匙
糖、胡椒粉、油、麻油及
紹酒各少許
生粉 2 茶匙

1/2 tsp salt
2 tsps light soy sauce
sugar
ground white pepper
oil
sesame oil
Shaoxing wine
2 tsps caltrop starch

做法 Method

1. 雞洗淨，抹乾水分，斬件。
2. 冬菇及金華火腿切片；大頭菜用水略浸，切片。
3. 將所有材料放入深碟內，拌入調味料，醃約半小時。
4. 燒熱水，隔水蒸約 15 分鐘即可品嘗。

1. Wash the chicken. Wipe dry. Chop into pieces.
2. Slice the black mushrooms and Jinhua ham. Soak the preserved mustard root in water for a while. Slice.
3. Put all the ingredients on a deep dish. Mix in the seasoning. Rest for about 1/2 hour.
4. Bring water to the boil. Steam the chicken for about 15 minutes. Serve.

潮式椒醬肉

豬肉 3 兩
三色甜椒各半個
炸脆花生 2 兩
甜菜甫 3 兩
豆腐乾 2 件
蝦米 1 兩
磨豉醬 1 湯匙（加水 1 湯匙拌勻）
豆瓣醬 1 湯匙
蒜茸 1 茶匙
葱 2 條（切碎）

113 g pork
1/2 green bell pepper
1/2 red bell pepper
1/2 yellow bell pepper
75 g deep-fried peanuts
113 g sweet pickled turnip
2 pieces dried tofu
38 g dried shrimps
1 tbsp ground bean sauce
(mixed with 1 tbsp of water)
1 tbsp chilli bean sauce
1 tsp finely chopped garlic
2 sprigs spring onion (chopped)

老抽半湯匙
生抽 1 茶匙
糖 1 1/2 茶匙
鹽 1/3 茶匙
胡椒粉、麻油及水各少許

1/2 tbsp dark soy sauce
1 tsp light soy sauce
1 1/2 tsps sugar
1/3 tsp salt
ground white pepper
sesame oil
water

做法
Method

1. 豬肉洗淨，切粒；三色甜椒洗淨，去蒂及籽，切塊；豆腐乾切片；甜菜甫切粒。

2. 燒熱油 3 湯匙，下蒜茸、磨豉醬及豆瓣醬爆香，加入肉粒、蝦米、豆腐乾及菜甫拌炒，下甜椒及調味料煮至汁液收乾，灑上葱粒及炸花生炒勻，上碟，冷熱吃皆宜。

1. Wash and dice the pork. Rinse the bell peppers. Remove the stalks and seeds. Cut the bell peppers into pieces. Slice dried tofu and dice sweet pickled turnip.

2. Heat 3 tbsps of oil. Stir-fry the garlic, ground bean sauce and chilli bean sauce until fragrant. Add the pork, dried shrimps, dried tofu and pickled turnip. Stir-fry evenly. Put in the bell peppers and the seasoning. Cook until the sauce dries. Sprinkle with the spring onion and peanuts. Stir-fry evenly. Put on a plate. Serve hot or cold.

Stir-fried Pork with Chilli Bean Sauce in Chaozhou Style

小秘訣 Tips

- 磨豉醬較濃稠，先用水調稀，以免下鑊後容易焦燶。
- 蝦米毋須用水浸泡，略沖即可，可保留蝦米香氣。
- The ground bean sauce is a bit thick. To avoid it getting burnt, add water to make it thin before cooking.
- It is not necessary to soak the dried shrimp in water. Slightly rinsing it can help retain its aroma.

Cooking fundamentals: Stir-frying

入廚必修課：炒

即時睇片 ○┈┈┈▶

技巧重點

- 選用加熱面積較廣的中式鑊。
- 抹乾鑊，加熱後見冒煙下油，轉動鑊面令油平均分佈，材料才不會黏鑊。
- 各項材料切成大小相若，抹乾水分。
- 加入食材先後有序：先料頭，後肉類，最後下蔬菜。
- 用中小火炒料頭，令香氣滲入油內。
- 以大火短時間不停翻炒食材。
- 先將全部調味料拌勻，一併加入鑊內調味，令整道菜調味均勻。
- 沿鑊邊潷酒，別直接澆在食材上。

Key Techniques

- Choose a Chinese wok with a wide heating surface area.
- Dry the wok and heat it up. When smokes emerge, put in oil and swirl it around to ensure the oil evenly distributed to prevent the ingredients from sticking.
- Cut the ingredients into pieces in similar sizes and then dry them.
- Add the ingredients in order: Aromatics are added first, then the meat, and finally the vegetable
- Stir-fry the aromatics over low-medium heat to allow their fragrance to infuse into the oil.
- Keep stir-frying the ingredients over high heat for a short time.
- Mix all the seasoning well beforehand, and then put into the wok to season the dish evenly.
- Don't spatter cooking wine directly on the ingredients! Spatter it along the side of the wok.

砂窩三蔥雞焗蟹
Chicken and Crab with Mixed Onions in Casserole

材料
Ingredients

雞半隻
蟹 1 隻（約 12 兩）
京蔥 2 棵
洋蔥（小）1 個
蔥 6 條
薑 6 片
蒜肉 6 粒（切片）

1/2 chicken
1 crab (about 450 g)
2 stalks Peking scallion
1 small onion
6 sprigs spring onion
6 slices ginger
6 cloves skinned garlic (sliced)

醃料
Marinade

鹽、老抽及糖各半茶匙
生抽 1 茶匙
生粉、胡椒粉及麻油各少許

1/2 tsp salt
1/2 tsp dark soy sauce
1/2 tsp sugar
1 tsp light soy sauce
caltrop starch
ground white pepper
sesame oil

調味料
Seasoning

蠔油 1 湯匙
生抽及糖各半茶匙
生粉 1 茶匙
上湯 4 安士（約 115 毫升）

1 tbsp oyster sauce
1/2 tsp light soy sauce
1/2 tsp sugar
1 tsp caltrop starch
4 oz. (115 ml) stock

121

1. 雞洗淨，抹乾水分，斬件，加入醃料拌勻醃半小時。

2. 蟹劏好，去鰓、去奄，洗淨，斬件。

3. 京蔥及蔥洗淨，切短度；洋蔥去外衣，切條。

4. 燒熱鑊，下油5湯匙，加入薑片、蒜片、京蔥、洋蔥及蔥白爆香，下雞件爆炒至半熟，加入蟹件炒至蟹熟，潷酒，下調味料及青蔥段拌勻。

5. 燒熱瓦煲，掃上油，將全部材料轉放瓦煲內，煮片刻即可原煲上桌。

1. Wash the chicken. Wipe dry. Chop into pieces. Mix with the marinade and rest for 1/2 hour.

2. Gut the crab. Remove the gills and abdomen. Wash thoroughly. Chop into pieces.

3. Rinse the Peking scallion and spring onion. Cut into short sections. Skin the onion. Cut into strips.

4. Heat a wok. Put in 5 tbsps of oil. Stir-fry the ginger, garlic, Peking scallion, onion and white part of the spring onion until fragrant. Put in the chicken and stir-fry until rare. Add the crab and stir-fry until done. Sprinkle with wine. Mix in the seasoning and green part of the spring onion.

5. Heat a casserole. Brush with oil. Transfer all the ingredients into the casserole. Cook for a moment. Serve with the casserole.

小秘訣 Tips

這道餸的香味來自京葱、洋葱及葱，必須先爆香三葱，讓氣味徹底散發出來，才放入雞塊及蟹件拌炒。

The fragrance of this dish comes from the Peking scallion, onion and spring onion. They must be stir-fried to let the beautiful smell permeate before stir-frying with the chicken and crab.

蛋白脆雞翼
Deep-fried Chicken Wings in Egg Whites

材料 Ingredients

雞中翼 1 斤
蛋白 3 個
粟粉 4 至 5 湯匙

600 g chicken mid-joint wings
3 egg whites
4 to 5 tbsps corn flour

小秘訣 Tips

適當地控制油溫，將火力調至慢火炸雞翼，可嘗到嫩滑的蛋白，雞翼也容易熟透。

Manage the heat properly. Turn the heat down to deep-fry the chicken wings. You can then taste the creamy egg whites. It is also easy for the chicken wings to be cooked through.

醃料 Marinade

鹽及生抽各 1 茶匙
胡椒粉及麻油各少許
糖半茶匙
薑汁及紹酒各 1 湯匙

1 tsp salt
1 tsp light soy sauce
ground white pepper
sesame oil
1/2 tsp sugar
1 tbsp ginger juice
1 tbsp Shaoxing wine

做法 Method

1. 雞翼洗淨，抹乾水分，加入醃料拌勻醃約半小時。
2. 蛋白拂勻，加入粟粉輕拌成滑蛋白漿，下雞翼沾滿蛋白液。
3. 燒熱油，下雞翼用慢火炸至金黃香脆，再調大火略炸，即可上碟。

1. Rinse the chicken wings. Wipe dry. Mix with the marinade and rest for about 1/2 hour.
2. Whisk the egg whites. Gently fold in the corn flour until it is smooth. Put in the chicken wings and coat with the egg white mixture.
3. Heat the oil. Deep-fry the chicken wings over low heat until golden and crisp. Turn to high heat and slightly deep-fry. Put on a plate. Serve.

蝦醬煎豬扒

Fried Pork Chop with Shrimp Paste

材料
Ingredients

肥豬扒 12 兩
450 g fat pork chop

醃料
Marinade

蝦醬 2 湯匙
糖半茶匙
蒜茸粉 1 茶匙
紹酒 1 湯匙
油 1 茶匙
胡椒粉及麻油各少許

2 tbsps shrimp paste
1/2 tsp sugar
1 tsp garlic powder
1 tbsp Shaoxing wine
1 tsp oil
ground white pepper
sesame oil

做法
Method

1. 豬扒洗淨，抹乾水分，用刀背略剁鬆，下醃料拌勻醃約 1 小時。
2. 將豬扒表面多餘的蝦醬抹掉，撲上生粉，放入熱油內以中火煎至金黃色，上碟，瀝乾油分即可。

1. Wash the pork chop. Wipe dry. Chop with the back of a knife to loosen the meat texture. Mi with the marinade and rest for about 1 hour.
2. Wipe the excess shrimp paste off the pork chop. Coat with caltrop starch. Fry in oil ove medium heat until golden. Drain. Serve.

小秘訣 Tips

不同品牌的蝦醬鹹味各異，建議先試味才醃製豬扒，否則豬扒太鹹，浪費食材。

The salty taste of shrimp paste varies with different brands. It is better to try the taste before using it to marinate the pork chop. It is a waste of food if the pork chop is too salty.

豉椒帶子

Stir-fried Scallops with Fermented Black Beans and Bell Peppers

材料 Ingredients

帶子 1 斤
青、紅甜椒各 1 個
薑 2 片
蒜肉 3 粒（剁茸）
豆豉 1 1/2 湯匙
葱 3 條（切段）

600 g scallops
1 green bell pepper
1 red bell pepper
2 slices ginger
3 cloves skinned garlic
(finely chopped)
1 1/2 tbsps fermented black beans
3 sprigs spring onion
(cut into sections)

醃料 Marinade

薑汁 1 湯匙
鹽半茶匙
生粉 1 湯匙

1 tbsp ginger juice
1/2 tsp salt
1 tbsp caltrop starch

獻汁 Thickening sauce

鹽及糖各半茶匙
生粉 1 茶匙
胡椒粉及麻油各少許
水 3 湯匙
* 拌勻

1/2 tsp salt
1/2 tsp sugar
1 tsp caltrop starch
ground white pepper
sesame oil
3 tbsps water
* mixed well

小秘訣 Tips

想吃到爽滑不韌的帶子，記着以下兩點要訣：用大滾水飛水，短時間即盛起；泡油時動作快捷。

Note the two keys for tasting soft and crunchy scallops: Blanch with heavy boiling water for a short time and then dish up at once. Deep-fry in hot oil swiftly.

做法
Method

1. 青、紅甜椒洗淨，去蒂及籽、切角；豆豉洗去外皮。

2. 帶子解凍，洗淨，瀝乾水分，下醃料拌勻醃半小時，放入滾水煮約 2 分鐘，過冷河，抹乾水分，再泡熱油 20 秒，盛起。

3. 燒熱油 3 湯匙，下甜椒拌炒，灑入水半湯匙炒熟，盛起。

4. 燒熱油 3 湯匙，下蒜茸、豆豉及薑片爆香，放入帶子大火略炒，濽酒，下甜椒及葱段炒勻，傾入獻汁略炒即成。

1. Rinse the green and red bell peppers. Remove the stalks and seeds. Cut into triangles. Wash away the skin of the fermented black beans.

2. Defrost the scallops. Rinse and drain. Mix with the marinade and rest for 1/2 hour. Put into boiling water and blanch for about 2 minutes. Rinse in cold water. Wipe dry. Deep-fry in hot oil for 20 seconds. Dish up.

3. Heat 3 tbsps of oil. Stir-fry the bell peppers. Drizzle with 1/2 tbsp of water. Stir-fry until done. Set aside.

4. Heat 3 tbsps of oil. Stir-fry the garlic, fermented black beans and ginger until fragrant. Put in the scallops. Roughly stir-fry over high heat. Sprinkle with wine. Add the bell peppers and spring onion. Stir-fry evenly. Pour in the thickening sauce. Roughly stir-fry and serve.

酥炸雞肉餅
Deep-fried Chicken Patties

材料
Ingredients

雞柳肉 450 克
西芹粒、洋葱粒及甘筍粒各 3 湯匙
雞蛋 2 個（拂勻）
麵粉及麵包糠各半杯

450 g chicken fillets
3 tbsps diced celery
3 tbsps diced onion
3 tbsps diced carrot
2 eggs (whisked)
1/2 cup flour
1/2 cup breadcrumbs

調味料
Seasoning

鹽 1 茶匙
生抽 2 茶匙
生粉 2 湯匙
蛋液半個
薑汁、紹酒、蒜鹽、胡椒粉及麻油各少許

1 tsp salt
2 tsps light soy sauce
2 tbsps caltrop starch
1/2 egg wash
ginger juice
Shaoxing wine
garlic salt
ground white pepper
sesame oil

做法
Method

1. 西芹粒、洋葱粒及甘筍粒用少許油炒香，下鹽調味炒勻，盛起備用。

2. 雞柳肉略沖淨，抹乾水分，去掉肉筋，剁碎，加入調味料拌勻成糰，分成 12 等份，每份包入蔬菜粒。

3. 雞肉餅輕輕依序沾上麵粉、蛋液及麵包糠，隨即放入熱油內，炸至表面呈微金黃色及外皮酥脆，瀝乾油分，伴茄汁蘸吃。

1. Stir-fry the celery, onion and carrot with a little oil until aromatic. Season with salt. Stir-fry evenly. Set aside.

2. Slightly rinse the chicken fillets. Wipe dry. Remove the tendons and chop up. Mix with the marinade evenly to become chicken dough. Divide into 12 equal parts. Stuff each part with diced vegetables.

3. Dip the chicken patty in flour, egg wash and breadcrumbs in order. Immediately put into hot oil. Deep-fry until the surface is light brown and crisp. Drain. Serve with ketchup.

魚乾烤茄子
Roasted Eggplants with
Dried Plaice

小秘訣 Tips

原條大地魚乾烘乾後，撕出魚肉並壓碎，儲存雪櫃隨時使用。雜貨店有現成的大地魚粉出售。

Toast the whole dried plaice and then tear the meat off. Crush the meat and keep it in a refrigerator for use at any time. Ready-made products are available in groceries.

材料 Ingredients

茄子 1 斤
大地魚乾碎 4 湯匙
葱粒 2 湯匙
蒜茸 1 湯匙

600 g eggplants
4 tbsps chopped dried plaice
2 tbsps diced spring onions
1 tbsp finely chopped garlic

調味料 Seasoning

老抽及生抽各 2 湯匙
糖、上湯及麻油各 1 湯匙
胡椒粉少許

2 tbsps dark soy sauce
2 tbsps light soy sauce
1 tbsp sugar
1 tbsp stock
1 tbsp sesame oil
ground white pepper

做法 Method

1. 預熱焗爐，放入茄子烘烤至外皮呈焦黑色，取出，隨即放入已煲煮的冰水內略浸，撕掉外皮，抹乾水分。
2. 茄子去蒂，切成 3 吋長段，再切條，上碟，鋪上魚乾。
3. 燒熱鑊，下油 3 湯匙，下蒜茸略炒，澆在茄子上。
4. 再燒熱調味料，澆於茄子，趁熱享用。

1. Preheat an oven. Bake the eggplants until the skin turns burned black. Take out. Immediately soak in ice water for a while. Tear the skin off. Wipe dry.
2. Remove the stalks of the eggplants. Cut into 3-inch long sections. Cut into strips. Dish up. Arrange the dried plaice on top.
3. Heat a wok. Put in 3 tbsps of oil. Roughly stir-fry the garlic. Sprinkle on the eggplants.
4. Heat the seasoning. Sprinkle on the eggplants. Serve hot.

鯪魚脊半斤
雞蛋 6 個
冬菇 8 個
（處理方法參考 p.12，切條）
生菜半斤
蒜茸半茶匙

300 g mud carp (spinal part)
6 eggs
8 dried black mushrooms
 (refer to p.12 for method;
shredded)
300 g lettuce
1/2 tsp finely chopped garlic

調味料
Seasoning

鹽及生粉各 1 茶匙
胡椒粉少許

1 tsp salt
1 tsp caltrop starch
ground white pepper

獻汁
Thickening sauce

蠔油 2 湯匙
生抽、老抽、糖及生粉
各 1 茶匙
麻油半茶匙
水 5 湯匙
* 拌勻

2 tbsps oyster sauce
1 tsp light soy sauce
1 tsp dark soy sauce
1 tsp sugar
1 tsp caltrop starch
1/2 tsp sesame oil
5 tbsps water
* mixed well

碧綠扒魚腐
Fish Curd on Greens

做法
Method

1. 鯪魚脊放在砧板（皮向下），用刀刮出魚肉，略剁，盛起，下調味料用筷子順方向拌至起膠，加入蛋 1 個，順一方向續拌至起膠，重複此步驟，直至加入全部雞蛋為止。
2. 燒熱鑊下油（油溫不要太高），舀入鯪魚蛋漿，待魚腐漲大及浮面，立即盛起，瀝乾油分。
3. 生菜放入油鹽水內灼至八成熟，上碟備用。
4. 燒熱鑊，下油 2 湯匙，爆香蒜茸，加入魚腐及冬菇快炒，潷酒，埋獻略煮，澆在生菜上即可。

1. Put the mud carp on a chopping board (skin down). Scrape off the meat. Chop gently. Add the seasoning. Evenly stir with chopsticks in one direction until gluey. Add 1 egg. Stir again in one direction until gluey. Repeat the steps until all the eggs are added.
2. Heat a wok. Put in oil (the oil temperature need not be too high). Spoon in the mud carp mixture. When the deep-fried fish curd plumps out and floats, dish up immediately. Drain.
3. Blanch the lettuce in water with oil and salt until 80% done. Put on a plate. Set aside.
4. Heat a wok. Put in 2 tbsps of oil. Stir-fry the garlic until fragrant. Add the fish curd and black mushrooms. Stir-fry swiftly. Sprinkle with wine. Mix in the thickening sauce. Cook for a while. Pour onto the lettuce. Serve.

小秘訣 Tips

見魚腐浮面即盛起，別炸太久，否則欠鬆軟質感。

When the fish curd floats, dish up at once. Deep-frying for too long will make it less fluffy.

金菇花枝卷

Steamed Enokitake Mushroom Cuttlefish Rol

小秘訣 Tips

將墨魚切薄片時，盡量薄一點，以免捲起時太厚，不易成功之餘，賣相也不美。

Slice the cuttlefish as thin as possible; otherwise, it will be too thick after rolling up. Too thick a roll will give a poor presentation and the dish can hardly be made successful.

136

材料
Ingredients

墨魚 1 隻（約 1 1/2 斤）
燒鴨 1/4 隻
西芹 2 兩
金菇 4 兩
蟹柳 8 條

1 cuttlefish (about 900 g)
1/4 roast duck
75 g celery
150 g Enokitake mushrooms
8 imitated crab sticks

醃料
Marinade

薑汁 1 湯匙
鹽、胡椒粉、麻油及生粉各半茶匙

1 tbsp ginger juice
1/2 tsp salt
1/2 tsp ground white pepper
1/2 tsp sesame oil
1/2 tsp caltrop starch

調味料
Seasoning

蠔油半湯匙
老抽及糖各半茶匙
生粉 1 茶匙
麻油半茶匙
上湯 112 毫升

1/2 tbsp oyster sauce
1/2 tsp dark soy sauce
1/2 tsp sugar
1 tsp caltrop starch
1/2 tsp sesame oil
112 ml stock

做法
Method

1. 墨魚劏好，洗淨，抹乾水分，墨魚肉片成薄片（2 吋濶 x 3 吋長），下醃料略醃。
2. 燒鴨去骨，取肉，切成條狀。
3. 西芹撕去老筋，切條；金菇切掉根部，略洗，兩者放入滾水內略灼，盛起備用。
4. 墨魚片鋪平，放上燒鴨、西芹、金菇及蟹柳，輕輕捲成條狀，以生粉封口，隔水大火蒸 4 分鐘。
5. 燒熱少許油，下調味料煮滾成薄獻汁，澆在花枝卷上即可。

1. Gut the cuttlefish. Wash thoroughly and wipe dry. Cut into fine pieces (2-inch wide x 3-inch long). Roughly mix with the marinade.
2. Bone the roast duck. Take the meat. Cut into strips.
3. Tear the tough strings off the celery. Cut into strips. Cut away the roots of the Enokitake mushrooms. Slightly rinse. Put the above ingredients in boiling water and blanch for a while. Set aside.
4. Lay the cuttlefish flat. Put on the roast duck, celery, Enokitake mushrooms and crab sticks. Gently roll up into a strip. Seal the opening with caltrop starch. Steam over high heat for 4 minutes.
5. Heat a little oil. Put in the seasoning and bring to the boil. Sprinkle over the cuttlefish rolls. Serve.

叉燒醬焗雞
Chicken in BBQ Pork Sauce

材料
Ingredients

雞半隻
蜜糖 1 1/2 湯匙
葱 2 條（切粒）

1/2 chicken
1 1/2 tbsps honey
2 sprigs spring onion (diced)

醃料
Marinade

叉燒醬 2 湯匙
薑汁 1 湯匙
蒜茸粉 1 茶匙
生抽半湯匙
老抽 1 湯匙
糖 1 茶匙
鹽 1/4 茶匙
胡椒粉及麻油各少許

2 tbsps BBQ pork sauce
1 tbsp ginger juice
1 tsp garlic powder
1/2 tbsp light soy sauce
1 tbsp dark soy sauce
1 tsp sugar
1/4 tsp salt
ground white pepper
sesame oil

做法
Method

1. 雞洗淨，抹乾水分，斬件，下醃料拌勻醃 1 小時。
2. 燒熱鑊，下油 4 湯匙，放入雞件用中火爆香，不斷翻炒，潷酒，調至慢火，加蓋焗片刻再拌炒約 10 分鐘，加入蜜糖拌勻，灑上葱花即成。

1. Wash the chicken. Wipe dry. Chop into pieces. Mix with the marinade and rest for 1 hour.
2. Heat a wok. Put in 4 tbsps of oil. Stir-fry the chicken over medium heat until fragrant. Ke stir-frying. Sprinkle with wine. Turn to low heat. Put a lid on and rest for a while. Stir-fry aga for about 10 minutes. Mix in the honey. Sprinkle with the spring onion. Serve.

小秘訣 Tips

因蜜糖容易搶火，宜上碟前才拌入，以免太早加入容易燒焦。

It is easy for the honey to get burnt if heated for too long. It is better to add the honey just before serving.

越式咖喱煎鯧魚

Vietnam-style Curry Pomfret

小秘訣 Tips

此餸最適合以鷹鯧魚烹調，但價錢略貴；也可選用燕子鯧代替。

Chinese pomfret is most suitable for making this dish, but it is a bit expensive.
You may use swallowtail pomfret instead.

材料
Ingredients

鯧魚 1 條（約 12 兩）
香茅 1 枝
紅辣椒 1 隻
乾葱 4 粒
蒜肉 2 粒
咖喱粉 1 茶匙
麵粉 1 湯匙

1 pomfret (about 450 g)
1 stalk lemongrass
1 red chilli
4 shallots
2 cloves skinned garlic
1 tsp curry powder
1 tbsp flour

醃料
Marinade

鹽半茶匙
胡椒粉少許
生粉 1 茶匙（後下）

1/2 tsp salt
ground white pepper
1 tsp caltrop starch (added last)

調味料
Seasoning

鹽 1/4 茶匙
糖及生抽各 1 茶匙
水 3/4 杯

1/4 tsp salt
1 tsp sugar
1 tsp light soy sauce
3/4 cup water

做法
Method

1. 鯧魚劏好，洗淨，在魚身表面斜剁數刀，抹上鹽及胡椒粉略醃。
2. 燒熱少許油，魚身撲上生粉，煎至兩面呈金黃色，上碟。
3. 香茅切碎；乾葱及蒜肉剁茸；紅椒切粒。
4. 燒熱油 2 湯匙，下香茅、乾葱及蒜茸爆香，放入咖喱粉及麵粉炒勻，下調味料拌煮，灒紹酒，下鯧魚略煮，灑入紅椒粒即可。

1. Gill and rinse the pomfret. Make a few scores on the fish body. Spread with salt and ground white pepper. Rest for a while.
2. Heat a little oil. Coat the fish with caltrop starch. Fry until both sides are golden. Put on a plate. Set aside.
3. Finely cut the lemongrass. Finely chop the shallots and garlic. Dice the red chilli.
4. Heat 2 tbsps of oil. Stir-fry the lemongrass, shallots and garlic until sweet-scented. Put in the curry powder and flour. Stir-fry evenly. Stir in the seasoning. Sprinkle with Shaoxing wine. Put in the fish and cook for a while. Sprinkle with the red chilli. Serve.

材料
Ingredients

蟹柳 6 條
雞蛋 7 個
忌廉芝士醬 1 杯

6 imitated crab sticks
7 eggs
1 cup cream cheese sauce

調味料
Seasoning

鹽半茶匙
粟粉 3 茶匙
清水 4 湯匙

1/2 tsp salt
3 tsps corn flour
4 tbsps water

做法
Method

1. 鍋內注入凍水，下雞蛋 3 個，開火煮約 10
   分鐘至熟，沖冷水，去殼，切碎，與忌廉芝
   士醬、鹽 1/4 茶匙及少許胡椒粉拌勻。

2. 雞蛋 4 個拂勻，加入調味料拌勻。平底鑊內
   燒熱少許油，下蛋液煎成 4 塊蛋皮。

3. 蛋皮鋪平，放上蟹柳條、熟蛋芝士醬，捲
   起，斜切成件享用。

1. Put cold water into a pot. Put in 3 eggs. Turn
   on heat and boil for about 10 minutes, or until
   done. Rinse with cold water. Remove the shell.
   Finely chop. Mix with the cream cheese sauce,
   1/4 tsp of salt and a little ground white pepper
   evenly.

2. Whisk 4 eggs. Stir in the seasoning. Heat a little
   oil in a pan. Put in the egg wash and fry to make
   4 pieces of egg skin.

3. Lay the egg skin flat. Put on the crab stick
   and cooked egg cheese sauce. Roll up. Cut
   diagonally into pieces. Serve.

蟹柳蛋卷
Crab Stick Egg Rolls

小秘訣 Tips

蛋液用水調稀，煎出來的蛋皮香軟嫩滑。

Make the egg wash thin by adding water. The fried egg skin will be soft and fragrant.

143

材料
Ingredients

冬瓜 8 至 9 斤
（以瓜蒂部份為宜）
瑤柱 2 粒
蝦仁 3 兩
冬菇 4 朵（用水浸軟）
金華火腿 1 兩
鴨腎 2 個
鮮蓮子 20 粒
瘦肉 2 兩
薑粒 2 湯匙
夜香花半兩（後下）
紗紙 1 張

4.8 kg to 5.4 kg winter melon
(choose the head part of the
winter melon)
2 dried scallops
113 g shelled shrimps
4 dried black mushrooms
(soaked in water to soften)
38 g Jinhua ham
2 dried duck gizzards
20 fresh lotus seeds
75 g lean pork
2 tbsps diced ginger
19 g Ye Xiang Hua (added last)
1 sheet mulberry paper

調味料
Seasoning

薑汁 2 湯匙
紹酒半茶匙
胡椒粉及鹽各適量

2 tbsps ginger juice
1/2 tsp Shaoxing wine
ground white pepper
salt

冬瓜盅
Winter Melon Soup

小秘訣 Tips

冬瓜要餘下一吋厚瓜肉，別挖得太薄，否則瓜
身容易穿爛。

The winter melon should have flesh of 1-inch
thick. Scooping out too much flesh will make
the winter melon thin. It will break easily.

做法
Method

1. 冬瓜去瓤及瓜肉，洗淨，瓜身邊沿切成花紋。
2. 瑤柱用水浸 2 小時，瑤柱水留用；金華火腿洗淨，飛水，切粒。
3. 燒滾水 3 碗，下金華火腿及瑤柱水煲約 10 分鐘，即成上湯。
4. 瘦肉及鴨腎切粒，與蝦仁一同飛水；蓮子飛水。
5. 所有材料（夜香花除外）與調味料拌勻，放入冬瓜盅內，傾入步驟 3 的火腿瑤柱上湯約七成滿，蓋上紗紙燉 3 小時，最後撒上夜香花即成。

1. Remove the pith and flesh of the winter melon. Rinse well. Carve the edge of the winter melon with artistic patterns.

2. Soak the dried scallops in water for 2 hours. Keep the dried scallop water. Rinse the Jinhua ham. Scald in boiling water and dice.

3. Bring 3 bowls of water to the boil. Put in the Jinhua ham and dried scallop water. Cook for about 10 minutes. The stock is done.

4. Dice lean pork and duck gizzards. Scald them and shrimps in the boiling water. Scald the lotus seeds.

5. Combine all the ingredients (except Ye Xiang Hua) with the seasoning. Put into the winter melon. Pour in the ham and scallop stock from step 3 until winter melon is 70% full. Cover with mulberry paper. Double-steam for 3 hours. Finally sprinkle with Ye Xiang Hua. Serve.

煎釀鯪魚
Fried Stuffed Mud Carp

材料 Ingredients

鯪魚 1 條（約 12 兩）
鯪魚肉 3 兩（剁爛）
冬菇 2 朵（處理方法參考 p.12；切粒）
臘肉粒 2 湯匙
蝦米 1 湯匙（剁碎）
葱 2 條（切粒）
芫茜 2 棵（切碎）
炸花生碎 1 湯匙

1 mud carp (about 450 g)
113 g mud carp meat
(chopped into puree)
2 dried black mushrooms
(refer to p.12 for method; diced)
2 tbsps diced preserved pork
1 tbsp dried shrimps
(finely chopped)
2 sprigs spring onion (diced)
2 stalks coriander
(finely chopped)
1 tbsp deep-fried peanuts (crushed)

調味料 Seasoning

鹽 3/4 茶匙
粟粉 1 湯匙
生抽 1 茶匙
糖半茶匙
胡椒粉少許
水 3 湯匙

3/4 tsp salt
1 tbsp corn flour
1 tsp light soy sauce
1/2 tsp sugar
ground white pepper
3 tbsps water

獻汁 Thickening glaze

鹽 1/4 茶匙
粟粉 1 茶匙
生抽 2 茶匙
糖 1 茶匙
水 4 湯匙
麻油少許

1/4 tsp salt
1 tsp corn flour
2 tsps light soy sauce
1 tsp sugar
4 tbsps water
sesame oil

小秘訣 Tips

- 購買鯪魚時，魚販可代起肉。
- 剁鯪魚肉時加入少許鹽，令肉質更爽。
- 要將全部鯪魚餡料一次過釀入鯪魚皮內，若分開兩次釀入，餡料容易脫落。
- The fishmonger can take the meat out for you when buying the mud carp.
- Chop the mud carp meat with a little salt. The meat will be crunchier.
- All the mud carp mixture must be filled inside the mud carp skin in one go. It is easy for the stuffing to fall apart if it is done separately in two times.

做法
Method

1. 鯪魚劏好，洗淨，抹乾水分，鯪魚起肉，鯪魚皮保留。

2. 原條鯪魚肉及3兩鯪魚肉盡量切成薄片，剁茸。

3. 冬菇、臘肉、蝦米、葱粒、芫茜、花生粒與鯪魚茸拌勻，加入調味料順一方向攪拌，逐少加入水拌勻，撻至起膠，冷藏片刻。

4. 鯪魚皮內腔塗抹少許生粉，釀入鯪魚肉，按實，魚身抹上生粉。

5. 燒熱鑊，下油半碗，放入釀鯪魚用慢火半煎炸至兩面金黃色，切件上碟。

6. 燒熱油1湯匙，潽酒，下獻汁煮滾，澆在魚身上即可。

1. Gill the mud carp and rinse well. Wipe dry. Take the meat out. Keep the whole mud carp skin.

2. Cut the whole mud carp meat and the 113 g mud carp meat as thin as possible. Chop into puree.

3. Combine the black mushrooms, preserved pork, dried shrimps, spring onion, coriander, peanuts and minced mud carp meat together. Add the seasoning. Stir in one direction evenly. Stir in water bit by bit. Throw into a bowl repeatedly until gluey. Chill for a moment.

4. Spread a little caltrop starch inside the mud carp skin. Stuff with the mud carp meat mixture. Press firmly. Spread caltrop starch on the fish.

5. Heat a wok. Put in 1/2 bowl of oil. Cook the mud carp over low heat by shallow to deep-frying. When both sides turn golden, dish up. Cut into pieces.

6. Heat 1 tbsp of oil. Sprinkle with wine. Put in the thickening glaze. Bring to the boil and pour onto the fish. Serve.

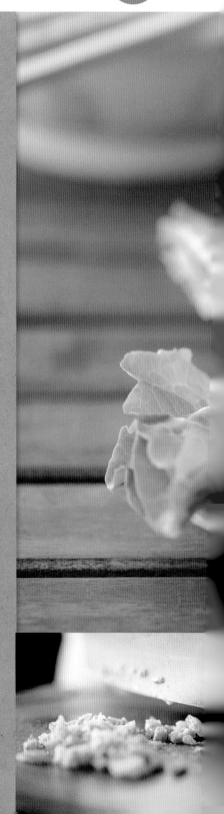

脆皮炸子雞
Crispy Deep-fried Chicken

小秘訣 Tips

雞皮多次塗抹麥芽糖上皮料，並徹底吹乾，炸後的雞皮才香脆。

Brush the maltose syrup mixture onto the chicken skin many times, and then air-dry it completely. The deep-fried chicken skin will be fragrant and crunchy.

材料
Ingredients

雞 1 隻（約 2 斤）
滷水料 1 包（花椒、八角、丁香、
草果、桂皮、甘草、陳皮）
蝦片 1 小包
1 chicken (about 1.2 kg)
1 packet Chinese marinade
(Sichuan peppercorns, star anises, cloves,
cardamom, cinnamon, liquorice, dried
tangerine peel)
1 small packet prawn crackers

上皮料
Ingredients for colouring skin

麥芽糖 1 茶匙（與滾水 2 湯匙拌勻）
浙醋、白醋及紹酒各 1 茶匙
生粉 1 湯匙
* 拌勻

1 tsp maltose syrup
(mixed with 2 tbsps boiling water)
1 tsp red vinegar
1 tsp white vinegar
1 tsp Shaoxing wine
1 tbsp caltrop starch
* mixed well

白滷水
White marinade

紹酒 4 湯匙
糖 8 茶匙
鹽 12 湯匙
水 16 杯

4 tbsps Shaoxing wine
8 tsps sugar
12 tbsps salt
16 cups water

做法
Method

1. 全部滷水料放入魚袋內，與白滷水材料煲半小時。
2. 雞劏好，洗淨，放入滷水汁內浸熟，取起並再放入煲內，重複數次，待滷水汁再滾起，關火，浸 15 分鐘，盛起。
3. 用上皮料均勻地塗抹雞皮兩次，吹乾約 4 小時。
4. 燒滾油，放入雞炸至金黃色，待涼，斬件上碟，伴蝦片享用。

1. Put all the ingredients of the Chinese marinade into a muslin bag. Cook with the white marinade in a pot for 1/2 hour.
2. Gut the chicken and wash thoroughly. Soak in the marinade until done. Lift up and put into the pot again. Repeat several times. When the marinade boils again, turn off heat. Soak the chicken for 15 minutes. Dish up.
3. Brush the maltose syrup mixture on the chicken skin evenly twice. Air-dry for about 4 hours.
4. Heat oil until it is scorching. Put in the chicken and deep-fry until golden. Let it cool down. Chop into pieces. Serve with the prawn crackers.

黃鱔 1 斤
銀芽 4 兩
韭黃 1 兩（切短度）
薑米 1 茶匙
蒜肉 2 粒（剁茸）
600 g yellow eel
150 g mung bean sprouts
(both ends removed)
38 g yellow chives
(cut into short sections)
1 tsp finely chopped ginger
2 cloves skinned garlic
(finely chopped)

醃料
Marinade

鹽 1/4 茶匙，生粉 3 茶匙
胡椒粉少許
1/4 tsp salt, 3 tsps caltrop starch
ground white pepper

調味料
Seasoning

老抽及蠔油各 1 湯匙
糖及麻油各 1 茶匙
鹽 1/4 茶匙
1 tbsp dark soy sauce
1 tbsp oyster sauce
1 tsp sugar
1 tsp sesame oil
1/4 tsp salt

生粉水
Caltrop starch
solution

生粉 1 1/2 茶匙，水 3 湯匙
* 拌勻
1 1/2 tsps caltrop starch
3 tbsps water
* mixed well

炒鱔糊
Stir-fried Shredded Eel in Sauce

做法
Method

1. 黃鱔放入滾水內飛水，過冷河，用刀刮去潺液，洗淨，抹乾水分。
2. 黃鱔起肉、去骨，切 1 1/2 吋長段，再切條，下醃料拌勻，泡油備用。
3. 韭黃與銀芽用少許鹽及油炒熟，盛起備用。
4. 燒熱油 2 湯匙，爆香薑米、蒜茸及黃鱔，濽酒，下調味料用大火炒勻，加入韭黃及銀芽拌炒，逐少加入生粉水炒勻，上碟即成。

1. Scald the eel in the boiling water. Rinse in cold water. Scrape off the skin mucus with a knife. Rinse well and wipe dry.
2. Bone the eel. Cut the meat into sections of 1 1/2 inches long. Cut into strips. Mix with the marinade evenly. Deep-fry in hot oil for a while. Set aside.
3. Stir-fry the yellow chives and mung bean sprouts with a little salt and oil until done. Set aside.
4. Heat 2 tbsps of oil. Stir-fry the ginger, garlic and eel until fragrant. Sprinkle with wine. Add the seasoning. Stir-fry evenly over high heat. Put in the yellow chives and mung bean sprouts and then stir-fry. Put in the caltrop starch solution bit by bit. Stir-fry evenly. Put on a plate. Serve.

小秘訣 Tips

- 魚販可代處理黃鱔、起骨及去潺液，不必費心。

- 黃鱔宜快炒，大火炒數下後，即加入配料拌勻即可。

- 以下是鱔糊的傳統吃法：上碟後在中央預留位置，放入蒜茸及滾油2湯匙，拌勻享用，蒜味濃郁。

- The fishmonger can help you handle and bone the eel. It comes easy.

- The eel needs to be stir-fried quickly. Roughly stir-fry and then add in other ingredients.

- The traditional way to enjoy stir-fried eel: After dishing up, make some room in the middle. Put in the finely chopped garlic and sprinkle 2 tbsps of boiling oil on top. Mix well. The dish will have an intense garlic flavour.

蝦膠釀竹笙
Stuffed Bamboo Fungus
with Minced Shrimp

材料
Ingredients

竹笙半兩
蝦肉 1 斤（蝦膠處理方法參考
肥豬肉 1 湯匙（煮熟、切粒）
蟹子 1 湯匙
西蘭花（小）1 個

19 g bamboo fungus
600 g shelled shrimps
 (refer to p.14 for method)
1 tbsp fat pork (cooked; diced)
1 tbsp crab roe
1 small broccoli

煨料
Cooking sauce

上湯 2 杯
葱 1 條
薑 2 片

2 cups stock
1 sprig spring onion
2 slices ginger

小秘訣 Tips

蝦膠用蛋白略醃，蝦肉更加滑嫩。

To marinate the minced shrimp with egg white will make
the meat smoother.

調味料
Seasoning

鹽半茶匙
生粉 1 1/2 茶匙
蛋白 1 湯匙
胡椒粉及麻油各少許
1/2 tsp salt
1 1/2 tsps caltrop starch
1 tbsp egg white
ground white pepper
sesame oil

獻汁
Thickening glaze

生粉 1 湯匙
鹽 1/4 茶匙
上湯 4 安士（約 115 毫升）
糖、胡椒粉及麻油各少許
1 tbsp caltrop starch
1/4 tsp salt
4 oz. (115 ml) stock
sugar
ground white pepper
sesame oil

做法
Method

1. 竹笙浸軟，沖洗多次。燒熱少許油，下薑片及葱爆香，濽酒，加入上湯 2 杯及竹笙煮片刻，瀝乾水分，剪成 3 至 4 吋長段，開邊，備用。
2. 肥豬肉粒、蝦膠及調味料順一方向拌勻，撻至起膠，冷藏一會。
3. 竹笙鋪平，抹上生粉，放上蝦膠，捲成圓筒狀，排於碟上，撒上蟹子，灑入少許油，蒸 10 分鐘。
4. 燒熱鑊，下油 1 湯匙，加入獻汁煮熱，濽酒，澆在竹笙面，伴已灼熟的西蘭花即可。

1. Soak the bamboo fungus in water to soften. Rinse many times. Heat a little oil. Stir-fry the ginger and spring onion until fragrant. Sprinkle with wine. Add 2 cups of stock and the bamboo fungus. Cook for a moment. Drain well and cut into long sections of 3 to 4 inches long. Cut open the bamboo fungus. Set aside.
2. Give the fat pork, minced shrimp and seasoning a good stir in one direction. Throw into a bowl repeatedly until gluey. Chill for a while.
3. Lay the bamboo fungus flat. Brush the caltrop starch. Put the shrimp mixture on top. Roll into a cylinder. Arrange on a plate. Sprinkle with the crab roe. Drizzle with a little oil. Steam for 10 minutes.
4. Heat a wok. Put in 1 tbsp of oil. Add the thickening glaze. Bring to the boil. Sprinkle with wine. Pour over the bamboo fungus. Put the blanched broccoli on the side. Serve.

脆皮炸鮮奶
Deep-fried Milk

即時睇片 ○······▶

材料
Ingredients

鮮奶 236 毫升（1 瓶）
椰汁 6 安士（約 170 毫升）
粟粉 6 湯匙
鹽 1 茶匙
蛋白 2 個
236 ml milk (1 bottle)
6 oz. (170 ml) coconut milk
6 tbsps corn flour
1 tsp salt
2 egg whites

脆漿料
Batter Ingredients

麵粉 8 安士（約 225 克）
生粉 1 湯匙
發粉 2 茶匙
鹽半茶匙
水 168 毫升
油 4 湯匙（後下）
8 oz. (225 g) flour
1 tbsp caltrop starch
2 tsps baking powder
1/2 tsp salt
168 ml water
4 tbsps oil (added last)

不藏私必學料理

小秘訣 Tips

- 鮮奶逐少份量拌入粟粉內，以免出現粉粒狀。
- 上碟前，開大火略炸鮮奶，迫出油分。
- Fold the milk in the corn flour little by little to prevent it from getting lumpy.
- Deep-fry the milk squares over high heat for a while before serving to let the excess oil out.

做法
Method

1. 粟粉篩勻,放在大碗內,慢慢加入鮮奶及椰汁,再下鹽拌勻,傾入煲內煮滾至濃稠
   不斷攪動至滑。

2. 蛋白打至挺身,加入奶糊內拌勻,開火略煮,放入已沾水之容器內,待涼,冷藏凝固

3. 麵粉、生粉、發粉及鹽篩勻至大碗內,加水順一方向拌勻,傾入油冷藏半小時(不
   攪動),取出拌勻,再冷藏 15 分鐘。

4. 取出已凝固的蛋白奶糕,切成小方塊,撲上乾生粉,沾上脆漿料,即放入滾油內用
   火炸至金黃色即成。

1. Sieve the corn flour. Put into a big bowl. Gradually add the milk and coconut milk. Put in
   salt and mix well. Pour into a pot. Bring to the boil. Cook until it is thick. Keep stirring unt
   turns creamy.

2. Whisk the egg whites until stiff. Fold in the milk mixture. Turn on heat. Cook for a while.
   into a container damped with water. Let it cool down. Chill to set.

3. Sieve the flour, caltrop starch, baking powder and salt into a big bowl. Add the water. Gi
   good stir in one direction. Put in the oil and chill for 1/2 hour (no stirring). Take out. Gi
   good stir. Chill again for 15 minutes.

4. Take out the set milk mixture. Cut into small squares. Coat with caltrop starch. Dip into
   batter. Put into hot oil at once and deep-fry over low heat until golden. Serve.

南乳齋煲

Braised Mixed Vegetables with Fermented Tarocurd in Casserole

材料 Ingredients

冬菇 1 兩（處理方法參考 p.12）
金針 3 錢
雲耳 3 錢
髮菜半兩
紅棗 8 粒（去核）
銀杏 2 兩
甜竹 3 兩
炸枝竹 4 兩
豆腐泡 3 兩
荷蘭豆 3 兩
黃芽白 12 兩
慈菇 8 粒
炸麵筋 4 兩
粉絲 1 扎
南乳 4 湯匙（用水 3 湯匙調勻）
甘筍數片（切成花狀）

38 g dried black mushrooms
(refer to p.12 for method)
12 g dried lily flowers
12 g cloud ear fungus
19 g dried black moss
8 red dates (cores removed)
75 g gingkoes
113 g sweetened beancurd strip
150 g deep-fried beancurd stick
113 g beancurd puffs
113 g snow peas
450 g Peking cabbage
8 Chinese arrowheads
150 g deep-fried gluten
1 bundle mung bean vermicelli
4 tbsps fermented tarocurd
(mixed with 3 tbsps water)
carrot slices (cut into floral shape)

調味料 Seasoning

紹酒及薑汁各 1 湯匙
片糖 3/4 塊
鹽適量
水 5 杯

1 tbsp Shaoxing wine
1 tbsp ginger juice
3/4 slab sugar
salt
5 cups water

獻汁 Thickening glaze

老抽及生粉各 1 茶匙
水 2 湯匙

1 tsp dark soy sauce
1 tsp caltrop starch
2 tbsps water

小秘訣 Tips

甜竹不用清洗，因遇水後會爛透，只用乾布抹淨即可。

It is no need to wash the sweetened beancurd strip as it will break in water. Just clean it with a dry cloth.

做法
Method

1. 粉絲浸軟，切段，瀝乾水分；金針、雲耳及髮菜浸透，洗淨，擠乾水分。
2. 熱鑊下少許油，爆香薑片及葱，濳酒，下水2杯及髮菜灼5分鐘，瀝乾水分。
3. 枝竹飛水，切段；荷蘭豆撕去老筋，與銀杏分別飛水。
4. 甜竹切成一吋長，放入熱油內炸至硬身，盛起。
5. 燒熱鑊，下油5湯匙，爆香南乳，加入金針及雲耳炒勻，下紅棗、銀杏、冬菇、慈菇、炸麵筋、豆腐泡、甜竹、枝竹、髮菜及粉絲略炒，放入調味料煮約10分鐘。
6. 下黃芽白煮熟，埋獻，下荷蘭豆及甘筍炒勻，轉放瓦煲品嘗。

1. Soak the mung bean vermicelli in water to soften. Cut into sections and drain. Soak the dried lily flowers, cloud ear fungus and black moss in water to soften. Rinse well. Squeeze water out.

2. Heat a wok. Add a little oil. Stir-fry the ginger and spring onion until aromatic. Sprinkle with wine. Add 2 cups of water and the black moss. Blanch for 5 minutes. Drain.

3. Blanch the deep-fried beancurd stick. Cut into sections. Tear the tough strings off the snow peas. Blanch the snow peas and gingkoes separately.

4. Cut the sweetened beancurd strip into pieces of 1-inch long. Deep-fry in hot oil until stiff. Set aside.

5. Heat a wok. Add 5 tbsps of oil. Stir-fry the fermented tarocurd until fragrant. Put in the lily flowers and cloud ear fungus. Stir-fry evenly. Add the red dates, gingkoes, black mushrooms, arrowheads, gluten, beancurd puffs, sweetened beancurd strip, beancurd stick, black moss and bean vermicelli. Roughly stir-fry. Add the seasoning and cook for about 10 minutes.

6. Put in the Peking cabbage. Cook until done. Mix in the thickening glaze. Add the snow peas and carrot. Give a good stir. Transfer to a casserole. Serve.

北菇扒海參
Braised Sea Cucumber with Black Mushrooms

材料
Ingredients

急凍海參 16 兩（已浸發）
冬菇 3 兩
（處理方法參考 p.12）
西生菜 1 個
薑 2 片
葱 2 條
600 g frozen sea cucumber
(rehydrated)
113 g dried black mushrooms
(refer to p.12 for method)
1 iceberg lettuce
2 slices ginger
2 sprigs spring onion

調味料
Seasoning

蠔油 3 湯匙
鹽半茶匙
老抽半湯匙
胡椒粉及麻油各少許
3 tbsps oyster sauce
1/2 tsp salt
1/2 tbsp dark soy sauce
ground white pepper
sesame oil

生粉水
Caltrop starch solution

生粉 1 湯匙
水 3 湯匙
* 拌勻
1 tbsp caltrop starch
3 tbsps water
* mixed well

做法
Method

1. 海參解凍，放入滾水內，與薑、葱及少許紹酒煨煮片刻，沖洗，抹乾水分。
2. 西生菜洗淨，用油、鹽、水灼熟，盛起。
3. 燒熱鑊，下少許油，灒酒，加入冬菇、海參及調味料燜約 10 分鐘，拌勻，下生粉水埋獻，鋪在生菜上即可。

1. Defrost the sea cucumber. Put in boiling water. Cook with the ginger, spring onion and a little Shaoxing wine for a while. Rinse and wipe dry.
2. Rinse the iceberg lettuce. Blanch with oil and salt until done. Set aside.
3. Heat a wok. Put in a little oil. Sprinkle with wine. Add the black mushrooms, sea cucumber and seasoning. Simmer for 10 minutes. Mix well. Stir in the caltrop starch solution. Lay on top of the iceberg lettuce. Serve.

鉢仔焗魚腸
Fish Intestine with Eggs in Earthen Bowl

材料
Ingredients

鯇魚腸 3 副（處理方法參考 p.15）
雞蛋 4 個
油條半條
芫茜粒及葱粒 3 湯匙
3 sets grass carp intestine
 (refer to p.15 for method)
4 eggs
1/2 piece deep-fried dough stick
3 tbsps diced coriander and spring onion

調味料
Seasoning

鹽 1 茶匙
薑汁 1 茶匙
陳皮茸 1 茶匙
欖角 6 粒（切粒）
紹酒 2 茶匙
胡椒粉少許
1 tsp salt
1 tsp ginger juice
1 tsp finely chopped
dried tangerine peel
6 preserved black olives (diced)
2 tsps Shaoxing wine
ground white pepper

做法
Method

1. 魚腸整理妥當，與調味料拌勻略醃。
2. 油條剪成薄片，備用。
3. 雞蛋拂勻，放入瓦鉢內，下魚腸及油條拌勻，隔水蒸約 15 分鐘，
   取出，放在明火上以慢火烘至水分全部收乾及散發香氣，最後
   撒上芫茜、葱及胡椒粉享用。

1. Treat the fish intestine properly. Mix with the seasoning. Rest for a
   while.
2. Cut the deep-fried dough stick into fine slices with scissors. Set aside.
3. Whisk the eggs. Put into an earthen bowl. Mix in the fish intestine
   and deep-fried dough stick. Steam for about 15 minutes. Put the bowl
   on the flame. Toast over low heat until the liquid dries out and it
   smells fragrant. Finally sprinkle with the coriander, spring onion and
   ground white pepper. Serve.

小秘訣 Tips

- 鯇魚腸體型較大，清洗時較容易及方便。
- 鯇魚腸黏附很多油脂肥膏，清洗時小心剪掉。
- 也可將瓦缽置於焗爐內，用 220℃ 烘焗至表面金黃香脆。

- If the grass carp intestine is larger in size, it will be easier for cleaning.
- There is plenty of fat sticking to the grass carp intestine. Cut them off with scissors carefully while washing.
- You may put the earthen bowl into an oven, and bake at 220 ℃ until the surface is golden and crisp.

南乳豬手
Braised Pork Knuckle with Red Fermented Tarocurd

小秘訣 Tips

- 豬手的肉質偏瘦、多肉；豬腳則較肥、骨多，建議選用豬手燜煮，吃得健康。

- 購買豬手時，緊記請肉販燒掉豬皮上的小毛。

- Pork knuckle is lean and meaty compared with pork leg which is fat and bony. For healthy eating, the pork knuckle is recommended.

- Ask the butcher to burn away the tiny hairs on the pork skin.

材料
Ingredients

豬手 1 隻
花生 4 兩
南乳 3 湯匙
蒜茸 1 茶匙
薑 2 片
1 pork knuckle
150 g peanuts
3 tbsps fermented tarocurd
1 tsp finely chopped garlic
2 slices ginger

調味料
Seasoning

糖及老抽各 1 湯匙
八角 1 粒
水 5 杯
1 tbsp sugar
1 tbsp dark soy sauce
1 star anise
5 cups water

做法
Method

1. 花生用水浸 2 小時，放入滾水內煲至軟身，瀝乾水分，備用。
2. 南乳及水 2 湯匙拌勻。
3. 豬手斬件，放入滾水內，加入白醋 1 湯匙煮約 10 分鐘，沖淨。
4. 燒熱油 1 湯匙，爆香薑片及蒜茸，加入豬手爆炒片刻，下花生、南乳及調味料煮滾，調至小火，燜約 1 小時，關火，加蓋焗 1 小時，灑上鹽再燜至豬手軟腍即可。

1. Soak the peanuts in water for 2 hours. Cook in boiling water until tender. Drain well. Set aside.
2. Combine the fermented tarocurd with 2 tbsps of water.
3. Chop the pork knuckle into pieces. Put into boiling water. Cook with 1 tbsp of white vinegar for about 10 minutes. Rinse.
4. Heat 1 tbsp of oil. Stir-fry the ginger and garlic until aromatic. Add the pork knuckle and stir-fry for a while. Put in the peanuts, fermented tarocurd and seasoning. Bring to the boil. Turn down the heat. Simmer for about 1 hour. Turn off heat. Put a lid on and leave for 1 hour. Sprinkle with salt and simmer until the pork knuckle is tender. Serve.

京都燻魚

Deep-fried Fish Belly in Sweet and Sour Sauce

材料
Ingredients

鯇魚腩 1 斤
葱 2 條
薑 2 片

600 g grass carp belly
2 sprigs spring onion
2 slices ginger

醃料
Marinade

生抽 1 茶匙
老抽 1 茶匙
鹽半茶匙
胡椒粉少許

1 tsp light soy sauce
1 tsp dark soy sauce
1/2 tsp salt
ground white pepper

京都汁
Sweet and sour sauce

生抽 1 湯匙
茄汁 3 湯匙
鹽 1/4 茶匙
糖 6 湯匙
麻油少許
水 6 安士（約 170 毫升）

1 tbsp light soy sauce
3 tbsps ketchup
1/4 tsp salt
6 tbsps sugar
sesame oil
6 oz. (170 ml) water

做法
Method

1. 鯇魚腩去鱗及內膜，洗淨，抹乾水分，切成大塊，下醃料拌勻，並用保鮮紙包裹，以重物徹底壓實，待約 3 小時。

2. 燒熱半鍋滾油，放入魚塊炸至金黃色，盛起。

3. 燒熱少許油爆香薑片及葱，去掉薑葱，傾入京都汁用小火煮至汁液濃稠，下魚塊拌勻，上碟享用。

1. Remove the scales and gill of the grass carp belly. Rinse well and wipe dry. Cut into chunks. Mix with the marinade. Wrap them with cling wrap and put a heavy object on the fish. Press tightly. Leave for 3 hours.

2. Heat 1/2 pot of oil until scorching. Deep-fry the fish until golden. Set aside.

3. Heat a little oil. Stir-fry the ginger and spring onion until fragrant. Remove the ginger and spring onion. Pour in the sweet and sour sauce. Simmer until the sauce reduces. Add the fish and mix well. Serve.

小秘訣 Tips

以重物壓着魚塊（以一天為佳），炸出來的鯇魚腩肉質較實。

Put a heavy object on the fish and leave it for 1 day (the desirable time). The deep-fried grass carp belly will have a firmer meat texture.

入廚必修課：炸

技巧重點

- 選用小鍋子，用油量較少。
- 將筷子放入熱油內測試油溫，見小泡冒出代表油溫熱透。
- 食材的水分必須徹底擦淨。
- 食材從鍋邊慢慢滑入，以免翻油。
- 別一下子放入太多食材，令油溫下降。
- 炸肉類時使用中火。
- 炸魚蝦蟹等海產，要待油滾後才放入，再用大火炸透，以短時間完成。
- 炸海產後的食油帶異味，可放入蘿蔔、薑或蒜頭略滾，去掉油腥味。
- 上碟前，用大火迫出材料的油分。
- 不想浪費大量食油，改用半煎炸方式，效果相同。

Key techniques

- Use a small pot for deep-frying to save oil.
- Test the temperature of oil with chopsticks before deep-frying. When bubbles form around the chopsticks, the oil is hot through.
- The ingredients must be dried thoroughly.
- From the side of the pot, slowly slip the ingredients down to the oil to avoid spilling.
- Don't put in too much ingredients at a time as the oil temperature will drop.
- Deep-fry the meat over medium heat.
- For deep-frying seafood such as fish, shrimp and crab, wait until the oil is hot through, and the add the seafood and deep-fry it thoroughly over high heat. It should be finished in a short time
- The used oil carries an odd smell. To remove the greasy smell, put in pieces of radish, ginger or garlic and deep-fry them slightly.
- Before serving, deep-fry the food over high heat to expel the oil from the food.
- If you don't want to waste too much oil, try shallow to deep-frying. The results will be the same

蝦膠菊花卷

Stuffed Squid Rolls with Minced Shrimp

材料
Ingredients

魷魚筒（中型）3 隻
海蝦 1 斤（蝦膠處理方法參考 p.14）
肥豬肉 1 湯匙（煮熟、切粒）
蟹子 1/4 茶匙
清雞湯 1 罐

3 medium-sized squid tubes
600 g marine shrimps (refer to p.14 for method)
1 tbsp fat pork (cooked; diced)
1/4 tsp crab roe
1 can chicken stock

醃料
Marinade

鹽 3/4 茶匙
蛋白 1 湯匙
胡椒粉及麻油各少許
生粉 1 湯匙

3/4 tsp salt
1 tbsp egg white
ground white pepper
sesame oil
1 tbsp caltrop starch

獻汁
Thickening glaze

清雞湯 4 安士（約 115 毫升）
生粉 1 茶匙
蛋白 1 個（後下）

4 oz. (115 ml) chicken stock
1 tsp caltrop starch
1 egg white (added last)

小秘訣 Tips

勿選用游水活蝦製蝦膠，因鮮蝦較黏殼，
難以去掉外殼。

It is hard to remove the shell of fresh
shrimp as its flesh will stick to the shell.
Therefore, do not use live shrimp to make
minced shrimp.

做法 Method

1. 蝦膠、肥肉粒及醃料順一方向拌勻，撻至起膠及有黏性，冷藏備用。

2. 魷魚筒洗淨，抹乾內外，切約 1 吋濶長段，用剪刀將半吋濶魷魚筒剪成鬚狀，其餘部份勿剪斷。

3. 燒滾清雞湯，下魷魚圈略灼，即散開成菊花形狀，抹乾水分，在內環抹上生粉，釀入蝦膠，隔水蒸 8 分鐘至熟，以蟹子裝飾。

4. 煮滾獻汁，關火，加入蛋白、胡椒粉及麻油拌勻，澆在蝦膠菊花卷上即可。

1. Stir the minced shrimp, fat pork and marinade evenly in one direction. Throw into a bowl repeatedly until gluey. Chill and set aside.

2. Wash the squid tubes. Wipe the inside and outside dry. Cut into long sections of about 1-inch wide. Cut 1/2-inch wide of the squid tube into tentacle-liked strips with scissors. Do not cut off from the rest part of the squid tube.

3. Bring the chicken stock to the boil. Blanch the squid rings slightly. It will spread out, looking like a chrysanthemum. Drain well and wipe dry. Spread caltrop starch inside the ring. Stuff with the shrimp mixture. Steam over high heat for 8 minutes, or until done. Decorate with the crab roe.

4. Bring the thickening glaze to the boil. Turn off heat. Mix in the egg white, ground white pepper and sesame oil. Sprinkle over the stuffed squid rolls with minced shrimp. Serve.

沙爹粉絲蟹煲

Satay Crab and
Mung Bean Vermicelli Casserole

小秘訣 Tips

海鮮容易熟透，所以蟹件泡油時油鑊必須燒至大滾，快速略炸，否則
蟹件欠酥脆口感。

It is easy for the seafood to be cooked through. Therefore, the oil
used for deep-frying the crab must be very hot and the deep-frying
process must be quick. Otherwise, the crab will not taste crunchy.

材料
Ingredients

肉蟹 1 斤
粉絲 1 小包
沙爹醬 1 湯匙
紅辣椒 1 隻（切粒）
葱 2 條（切段）
薑片及乾葱片各 1 茶匙
600 g male mud crab
1 small packet mung bean
vermicelli
1 tbsp satay sauce
1 red chilli (diced)
2 sprigs spring onion (sectioned)
1 tsp ginger slices
1 tsp shallot slices

醃料
Marinade

胡椒粉少許
生粉適量（後下）
ground white pepper
caltrop starch (added last)

調味料
Seasoning

上湯 3 杯
鹽及糖各 1/4 茶匙
3 cups stock
1/4 tsp salt
1/4 tsp sugar

做法
Method

1. 粉絲用水浸軟，盛起，切約 2 吋長。
2. 肉蟹去鰓及污物，洗淨，斬件，瀝乾水分，下醃料拌勻，泡油備用。
3. 瓦煲燒熱調味料，放入粉絲煮滾，備用。
4. 取另一個鑊，燒熱油 3 湯匙，爆香薑片、乾葱片、沙爹醬及蟹件，放在粉絲內同煮，灑入紅椒粒及葱段，上桌享用。

1. Soak the mung bean vermicelli in water to soften. Dish up. Cut into about 2-inch long sections.
2. Remove the gill and dirt of the mud crab. Rinse well. Chop into pieces and drain. Mix with the marinade. Deep-fry in scorching oil for a moment. Set aside.
3. Heat the seasoning in a casserole. Put in the mung bean vermicelli. Bring to the boil. Set aside.
4. Heat 3 tbsps of oil in a wok. Stir-fry the ginger, shallot, satay sauce and mud crab until aromatic. Put into the casserole containing the mung bean vermicelli. Cook together for a while. Sprinkle with the red chilli and spring onion. Serve.

蛋白 5 至 6 個（約 1 杯）
皮蛋 2 個
鮮奶 4 安士（約 115 毫升）
凍開水 7 安士（約 200 毫升）
紗紙 1 張

5 to 6 egg whites (about 1 cup)
2 preserved duck's eggs
4 oz. (115 ml) milk
7 oz. (200 ml) cold drinking water
1 sheet mulberry paper

雪映松花
Steamed Preserved Duck's Egg
Creamy Egg Whites

蛋白調味料
Seasoning for egg whites

鹽 3/4 茶匙
白醋 1/4 茶匙
橄欖油 2 茶匙

3/4 tsp salt
1/4 tsp white vinegar
2 tsps olive oil

皮蛋調味料
Seasoning for preserved duck's egg

糖及麻油各 1/4 茶匙

1/4 tsp sugar
1/4 tsp sesame oil

做法
Method

1. 皮蛋去殼，洗淨，切粒，下調味料略拌，放入已油之碟內。
2. 蛋白、鮮奶、凍開水及蛋白調味料放入深碗內拂傾入盛有皮蛋的碟內，蓋上紗紙以中火蒸約 7 分關火，再焗 2 分鐘即可。

1. Shell the preserved duck's eggs. Rinse and cut into Roughly mix with the seasoning. Put on a greased pla
2. Put the egg whites, milk, cold drinking water an seasoning for egg whites into a deep bowl. Give a whisk. Pour into the plate containing the preserved eggs. Cover with mulberry paper. Steam over mediu for about 7 minutes. Turn off heat. Leave for 2 m with the lid on. Serve.

小秘訣 Tips

蒸蛋白時蓋上紗紙或碟，蛋白特別嫩滑。

The steamed egg whites are particularly soft and creamy by covering with mulberry paper or a plate while steaming.

薑芽醬瓜鴨
Stir-fried Duck with
Young Ginger and Pickled Cucumber

小秘訣 Tips

薑芽於夏天當造，質感爽嫩，味清香。若買不到，可用醃子薑代替，
醬園店有醃子薑及醬瓜出售。

Young ginger is in season in summer. It has a crunchy texture and
a light aroma. If fresh young ginger is not available, use the pickled
instead. Pickled young ginger and cucumber can be bought at shops
selling sauce and pickles.

材料 Ingredients

鮮鴨肉 6 兩
薑芽（子薑）2 兩
醬瓜 1 兩
紅辣椒（長型）2 隻
蛋白 1 個
薑茸及蒜茸各半茶匙

225 g fresh duck meat
75 g young ginger
38 g pickled cucumber
2 long-shaped red chillies
1 egg white
1/2 tsp finely chopped ginger
1/2 tsp finely chopped garlic

醃料 Marinade

薑汁及紹酒各半茶匙
生抽、胡椒粉及麻油各少許
1/2 tsp ginger juice
1/2 tsp Shaoxing wine
light soy sauce
ground white pepper
sesame oil

獻汁 Thickening glaze

茄汁、蠔油及生抽各 1 茶匙
糖 2 茶匙
生粉 1 茶匙
鹽 1/3 茶匙
* 拌勻
1 tsp ketchup
1 tsp oyster sauce
1 tsp light soy sauce
2 tsps sugar
1 tsp caltrop starch
1/3 tsp salt
* mixed well

做法 Method

1. 鴨肉切成薄片，加入蛋白、少許生粉及醃料拌勻略醃。
2. 醬瓜及薑芽洗淨，擠乾水分，切片；紅辣椒去核，切成角形。
3. 燒熱油，下鴨肉泡油至八成熟，盛起。
4. 燒熱油 1 湯匙，下蒜茸及薑茸爆香，加入薑芽、醬瓜及紅辣椒拌炒，再放入鴨肉略炒，埋獻即成。

1. Finely slice the duck meat. Mix in the egg white, a little caltrop starch and the seasoning. Rest for a while.
2. Rinse the pickled cucumber and young ginger. Squeeze water out. Cut into slices. Remove the seeds of the red chillies. Cut into triangles.
3. Heat oil. Deep-fry the duck meat until it is 80% done. Set aside.
4. Heat 1 tbsp of oil. Stir-fry the garlic and ginger until sweet-scented. Add the young ginger, pickled cucumber and red chillies. Stir-fry for a while. Put in the duck meat and roughly stir-fry. Mix in the thickening glaze. Serve.

生炒鴿鬆
Stir-fried Minced Pigeon

材料 Ingredients

乳鴿 1 隻
豬肉 3 兩
冬筍 3 兩
芹菜 2 棵
冬菇 2 朵（浸軟）
馬蹄 4 粒
欖仁半兩
薑茸 2 茶匙
蒜茸 1 茶匙
葱粒 1 湯匙
西生菜半斤

1 pigeon
113 g pork
113 g winter bamboo shoots
2 stalks Chinese celery
2 dried black mushrooms
(soaked in water to soften)
4 water chestnuts
19 g olive nuts
2 tsps finely chopped ginger
1 tsp finely chopped garlic
1 tbsp diced spring onion
300 g iceberg lettuce

醃料 Marinade

鹽、糖及生粉各半茶匙
胡椒粉及麻油各少許
水 2 湯匙

1/2 tsp salt
1/2 tsp sugar
1/2 tsp caltrop starch
ground white pepper
sesame oil
2 tbsps water

獻汁 Thickening glaze

蠔油 1 湯匙
生粉半湯匙
鹽、糖、老抽及水各半茶匙
胡椒粉及麻油各少許

1 tbsp oyster sauce
1/2 tbsp caltrop starch
1/2 tsp salt
1/2 tsp sugar
1/2 tsp dark soy sauce
1/2 tsp water
ground white pepper
sesame oil

小秘訣 Tips

- 乳鴿及肉粒泡油時，油溫毋須太高，略滾即可。
- 加入獻汁後，炒至水分收乾，否則整道餸水汪汪，難以包吃。
- The temperature of oil for deep-frying the pigeon and pork should not be too high. Put in when the oil comes to a simmer.
- After adding the thickening glaze, keep stir-frying until the sauce dries; otherwise, it will be soggy and hard to be wrapped.

做法
Method

1. 欖仁放入熱油用慢火炸脆，盛起備用。

2. 冬菇及芹菜分別切粒；冬筍飛水，切粒；馬蹄去皮，用刀略拍，剁碎。

3. 乳鴿起肉，與豬肉分別切粒，下醃料拌勻略醃。

4. 燒熱鑊，下適量油至略滾，放入乳鴿粒及肉粒泡嫩油，瀝乾油分。

5. 燒熱油，下薑茸及蒜茸起鑊，放入乳鴿、肉粒、冬筍、冬菇、馬蹄及芹菜拌炒，灒酒，下獻汁略炒，灑上葱粒及欖仁，以生菜片包吃。

1. Deep-fry the olive nuts in oil over low heat to make them crunchy. Set aside.

2. Dice the black mushrooms and Chinese celery separately. Blanch the winter bamboo shoots and cut into dices. Skin the water chestnuts. Slightly pound with a knife. Finely chop up.

3. Take the meat of the pigeon. Dice the pigeon meat and pork separately. Mix in the marinade. Rest for a while.

4. Heat a wok. Put in some oil. When it comes to a simmer, put in the diced pigeon and pork. Deep-fry for a while. Drain well.

5. Heat oil in the wok. Stir-fry the ginger and garlic until fragrant. Put in the diced pigeon, pork, winter bamboo shoots, black mushrooms, water chestnuts and Chinese celery. Stir-fry for a while. Sprinkle with wine. Add the thickening glaze and roughly stir-fry. Sprinkle with the spring onion and olive nuts. Wrap in pieces of the iceberg lettuce to eat.

大良炒鮮奶
Stir-fried Milk with Crab Meat

小秘訣 Tips

想吃到嫩滑的炒鮮奶，秘訣在於將鑊燒至冒煙，下油至微熱，加入鮮奶混合料快手炒勻至凝固即可。

The secret of making creamy stir-fried milk is to heat the wok until it smokes and then add in oil. When the oil is just warm, stir-fry the milk mixture quickly until it is set.

材料
Ingredients

鮮奶 236 毫升（1 瓶）
蛋白 6 個
蟹肉 6 湯匙
草菇 6 粒
金華火腿茸 2 湯匙
米粉 1/4 兩

236 ml milk (1 bottle)
6 egg whites
6 tbsps crab meat
6 straw mushrooms
2 tbsps finely chopped Jinhua ham
10 g rice noodles

調味料
Seasoning

鹽 1 茶匙
生粉 1 湯匙

1 tsp salt
1 tbsp caltrop starch

做法
Method

1. 蟹肉吸乾水分，下紹酒及胡椒粉各少許拌勻。
2. 燒熱油，關火，下米粉炸至鬆脆，瀝乾油分備用。
3. 草菇飛水，用乾布抹淨，切片。
4. 蛋白打至挺身，下鮮奶打勻，加入調味料、草菇及蟹肉拌勻。
5. 燒熱鑊，下油 6 湯匙至微熱，傾入已拌勻的蛋白鮮奶混合料，用小火不斷拌炒至凝固，盛於炸米粉上，灑入金華火腿茸，趁熱享用。

1. Sop up the moisture in the crab meat. Mix in a little Shaoxing wine and ground white pepper.
2. Heat oil. When it is hot, turn off heat. Deep-fry the rice noodles until fluffy and crisp. Drain the oil.
3. Scald and clean the straw mushrooms with a dry cloth. Cut into slices.
4. Whisk the egg whites until stiff. Whisk in the milk. Add the seasoning, straw mushrooms and crab meat. Mix up.
5. Heat a wok. Put in 6 tbsps of oil. When it is just warm, pour in the milk mixture. Keep stir-fry over low heat until set. Put on top of the deep-fried rice noodles. Sprinkle with the Jinhua ham. Serve hot.

百花琵琶豆腐

Deep-fried Tofu Mixed with Shrimp

小秘訣 Tips

- 豆腐略壓成茸後，才與蝦膠拌勻，可徹底混和。
- 先將瓦匙沾上熱油，令豆腐蝦膠容易脫離。

- Slightly press the tofu into puree so that it can be mixed with the minced shrimp entirely.
- Dip the earthen spoon in hot oil beforehand. The pear-shaped tofu mixture will come off easily.

材料
Ingredients

豆腐 2 塊
蝦膠 8 兩（處理方法參考 p.14）
雞蛋 3 個
薑米 1 茶匙
蒜茸半茶匙
葱花 3 湯匙

2 cubes tofu
300 g minced shrimp (refer to p.14 for method)
3 eggs
1 tsp finely chopped ginger
1/2 tsp finely chopped garlic
3 tbsps diced spring onion

調味料
Seasoning

鹽半茶匙
生粉 1 湯匙
蛋白 1 湯匙
胡椒粉及麻油各少許

1/2 tsp salt
1 tbsp caltrop starch
1 tbsp egg white
ground white pepper
sesame oil

獻汁
Thickening glaze

蠔油 1 湯匙
生抽 1 茶匙
老抽半茶匙
紹酒及生粉各 1 茶匙
清雞湯半杯

1 tbsp oyster sauce
1 tsp light soy sauce
1/2 tsp dark soy sauce
1 tsp Shaoxing wine
1 tsp caltrop starch
1/2 cup chicken stock

做法
Method

1. 蝦膠與調味料順一方向拌勻，撻至起膠及有黏性，加入豆腐及雞蛋略拌勻。
2. 燒滾油，瓦匙略沾上油至熱，用瓦匙舀起豆腐蝦膠放入滾油內，炸至微黃色，盛起，如琵琶形狀。
3. 燒熱少許油，下蒜茸及薑米爆香，加入獻汁煮滾，澆上琵琶豆腐，灑入葱花即成。

1. Stir the minced shrimp and seasoning evenly in one direction. Throw into a bowl repeatedly until sticky. Add the tofu and eggs and slightly mix up.
2. Heat oil until scorching. Slightly dip an earthen spoon in the oil to make it hot. Spoon the tofu mixture into the oil. Deep-fry until light brown. Dish up. It looks like a pear in shape.
3. Heat a little oil. Stir-fry the garlic and ginger until fragrant. Add the thickening glaze. Bring to the boil. Pour over the pear-shaped tofu. Sprinkle with the spring onion. Serve.

五香牛腱
Five-spice Beef Shin

到藥材店或雜貨店購買約 7 元滷水料即可；或市面有滷水料包出售，大部份香料已齊備。

Buy Chinese marinade ingredients at about $7 at Chinese herb shops or groceries. Packets of Chinese marinade ingredients are also available in the market. The packet has already included most of the spices.

材料
Ingredients

牛腱（金錢腱）1 斤
600 g beef shin

滷水料
Chinese marinade

陳皮半個
肉桂 2 片
香葉 4 片
草果 2 個
八角 2 粒
1/2 dried tangerine peel
2 slices cinnamon
4 bay leaves
2 cardamoms
2 star anises

白滷水
White marinade

葱 2 條
生抽及老抽各 5 湯匙
紹酒 3 湯匙
冰糖 1 1/2 湯匙
薑 2 片
水 6 杯
2 sprigs spring onion
5 tbsps light soy sauce
5 tbsps dark soy sauce
3 tbsps Shaoxing wine
1 1/2 tbsps rock sugar
2 slices ginger
6 cups water

做法
Method

1. 牛腱洗淨，飛水約 10 分鐘，過冷河，備用。
2. 滷水料用魚袋盛好，放入白滷水料內煮 10 分鐘，下牛腱煲 1 小時至軟身及熟透，預留滷水汁約 1 杯。
3. 將牛腱浸於滷水汁約 2 小時，取出，冷藏 2 至 3 小時，切成薄片，澆上麻油享用。

1. Rinse the beef shin. Blanch for about 10 minutes. Rinse in cold water. Set aside.
2. Put the Chinese marinade ingredients into a muslin bag. Put the bag in the white marinade. Cook for 10 minutes. Add the beef shin. Cook for 1 hour, or until it is tender and fully done. Reserve 1 cup of the cooked marinade.
3. Soak the beef shin in the marinade for about 2 hours. Take out. Chill for 2 to 3 hours. Finely slice and sprinkle with sesame oil. Serve.

絲瓜烙

Fried Angled Luffa with Duck Egg

材料 Ingredients

絲瓜 5 兩
鴨蛋 2 個（拂勻）
泰國番薯粉 6 湯匙
葱 2 條
芫茜 1 棵
水 4 湯匙

188 g angled luffa
2 duck eggs (whisked)
6 tbsps Thai sweet potato starch
2 sprigs spring onion
1 stalk coriander
4 tbsps water

調味料 Seasoning

鹽 1 茶匙
生抽半茶匙
胡椒粉及麻油各少許

1 tsp salt
1/2 tsp light soy sauce
ground white pepper
sesame oil

做法 Method

1. 絲瓜刨去皮，洗淨，切成細粒；芫茜及葱切小粒。
2. 番薯粉及水拌勻，下鴨蛋液略拌，加入調味料、絲瓜粒、芫茜及葱拌成蛋漿料。
3. 燒熱平底鑊，下油 6 湯匙，舀入蛋漿料煎至兩面金黃色，潷少許酒略煎即可。

1. Peel the angled luffa. Rinse and finely dice. Cut the coriander and spring onion into small dices.
2. Mix the sweet potato starch with water. Mix in the duck egg wash. Add the seasoning, angled luffa, coriander and spring onion. Give a good stir.
3. Heat a pan. Put in 6 tbsps of oil. Ladle in the egg mixture. Fry until both sides are golden. Sprinkle with a little wine and slightly fry. Serve.

小秘訣 Tips

- 泰國番薯粉較幼細，容易拌和，令蛋漿幼滑。
- 鴨蛋是這道餸之精髓，蛋味香濃，別用雞蛋代替。
- Having a delicate texture, Thai sweet potato starch can be easily combined with the egg wash to make it creamy.
- The essence of this dish is the duck egg which has an intense egg flavour. Do not replace it with chicken egg.

栗子燜雞
Simmered Chicken and Chestnut with Ground Bean Sauce

小秘訣 Tips

栗子肉放入滾水內煲5分鐘，在布上放數粒栗子，趁熱擦拭，容易去掉栗子外衣；切勿沖凍水。

Cook the shelled chestnuts in boiling water for 5 minutes. Then put a couple of the chestnuts on a cloth. Rub the skin while hot. It is easy for the skin to come off. Make sure do not rinse with cold water.

材料 Ingredients

光雞半隻（約 1 斤 4 兩）
栗子肉 8 兩
磨豉醬 2 茶匙
薑 3 片
乾葱肉 4 粒（切片）
蒜肉 4 粒
葱 2 條（切段）

1/2 chicken (about 750 g)
300 g shelled chestnuts
2 tsps ground bean sauce
3 slices ginger
4 shallots (sliced)
4 cloves garlic
2 sprigs spring onion
(cut into sections)

醃料 Marinade

生抽及紹酒各 2 茶匙
生粉半茶匙
胡椒粉少許

2 tsps light soy sauce
2 tsps Shaoxing wine
1/2 tsp caltrop starch
ground white pepper

獻汁 Thickening glaze

生粉 1 茶匙
水 2 湯匙
* 拌勻

1 tsp caltrop starch
2 tbsps water
* mixed well

調味料 Seasoning

鹽半茶匙
糖及生抽各 1 1/2 茶匙
老抽 1 茶匙

1/2 tsp salt
1 1/2 tsps sugar
1 1/2 tsps light soy sauce
1 tsp dark soy sauce

做法 Method

1. 雞洗淨，斬件，瀝乾水分，下醃料拌勻。
2. 栗子肉放入滾水內煲 5 分鐘，趁熱用布擦去外衣。
3. 燒熱鑊，下油 3 湯匙，爆香薑片、乾葱及蒜肉，加入雞件爆炒，下磨豉醬炒勻，潷酒，下水 56 毫升、調味料及栗子肉拌勻，加蓋燜 15 分鐘，埋獻，下葱段即成。

1. Wash the chicken. Chop into pieces and drain well. Mix with the marinade.
2. Cook the shelled chestnuts in boiling water for 5 minutes. Rub off the skin with a cloth while hot.
3. Heat a wok. Put in 3 tbsps of oil. Stir-fry the ginger, shallots and garlic until aromatic. Add the chicken and stir-fry. Put in the ground bean sauce. Stir-fry evenly. Sprinkle with wine. Add 56 ml of water, the seasoning and chestnuts. Mix up. Simmer with a lid on for 15 minutes. Stir in the thickening glaze. Put in the spring onion. Serve.

Cooking fundamentals: Stewing & Boiling
入廚必修課：燜煮

技巧重點

即時睇片 ○┅┅▶

- 蔬菜瓜類別切得太細碎。
- 先用大火燜煮，沸騰後將火下調續燜。
- 容易煮至散爛的食材稍後才放入。
- 煮肉時，見表面有浮沫，宜用密篩撇去。
- 水分不足時，加入熱水再燜煮。
- 調味料別一下子放太多，因燜煮後的味道會變得濃郁。
- 燜煮牛腩、牛腱、羊腩或鴨時，別太早加入鹽，會令肉質太鹹及太硬。
  宜燜煮焗透後，最後才下鹽調味。

Key techniques

- Don't cut the vegetables and gourds too finely.
- First heat the ingredients over high heat. When it comes to the boil, turn down the h[...]
  and slowly cook the ingredients.
- The ingredients that will easily break in cooking should be added at a later stage.
- Skim off the froth when boiling the meat.
- Top up with hot water when water is losing during stewing.
- The flavour will become intense after it is stewed, so try to avoid adding too m[...]
  seasoning at a time.
- Season the stewed beef brisket, beef shin, lamb belly or duck with salt after it is [...]
  done. Adding the salt too early will make the meat too salty and tough.

叉燒魚皮角

Fish Dumplings with BBQ Pork in Soup

材料
Ingredients

原條鯪魚肉 12 兩
叉燒 2 兩
冬菇 2 湯匙
（處理方法參考 p.12）
菜甫 1 湯匙
芹菜粒、甘筍粒及葱粒各 2 湯匙
韭黃 1 兩
芥蘭半斤

450 g whole mud carp meat
75 g BBQ pork
2 tbsps dried black mushroom
(refer to p.12 for method)
1 tbsp pickled radish
2 tbsps diced Chinese celery
2 tbsps diced carrot
2 tbsps diced spring onion
38 g yellow chives
300 g Chinese kale

上湯調味料
Ingredients for seasoned stock

上湯 5 杯
胡椒粉及麻油各少許

5 cups stock
ground white pepper
sesame oil

魚肉調味料
Seasoning for mud carp meat

鹽半茶匙
粟粉 1 湯匙
胡椒粉及水各少許

1/2 tsp salt
1 tbsp corn flour
ground white pepper
water

調味料
Seasoning

鹽、糖、生抽及麻油各半茶匙
生粉 1 茶匙
水 2 湯匙

1/2 tsp salt
1/2 tsp sugar
1/2 tsp light soy sauce
1/2 tsp sesame oil
1 tsp caltrop starch
2 tbsps water

小秘訣 Tips

• 以鯪魚肉代替雲吞皮，別捏得太厚，包入餡料後以免過度漲滿。

• 叉燒魚皮角可作為火鍋的配料或蒸吃。

• The mud carp meat mixture is used like the wanton wrapper. Knead it into thin pieces; otherwise, the dumplings will be overly plump.

• The fish dumplings with BBQ pork can be served in hot pots, or steamed.

做法
Method

1. 鯪魚肉洗淨，抹乾，用刀由上至下刮出魚肉，放入碗內，下粟粉、鹽、水及胡椒粉順一方向拌勻，撻至起膠。
2. 芥蘭放入滾水內，下油、糖及鹽略灼，盛起；其餘材料切成幼粒，備用。
3. 燒熱鑊，下冬菇、叉燒、芹菜、菜甫及甘筍爆香，加入調味料炒勻，灑入蔥粒，盛起。
4. 雙手沾上生粉，取鯪魚膠捏成扁圓形，包入叉燒餡料，摺成角形。
5. 燒滾上湯調味料，加入魚皮角煮滾，下芥蘭略煮，灑上韭黃即成。

1. Rinse the mud carp meat. Wipe dry. Scrape off the meat with a knife. Put into a bowl. Add the corn flour, salt, water and ground white pepper. Stir in one direction until even. Throw into the bowl repeatedly until sticky.
2. Blanch the Chinese kale in boiling water with oil, sugar and salt for a while. Set aside. Finely dice the rest ingredients. Set aside.
3. Heat a wok. Stir-fry the black mushroom, BBQ pork, Chinese celery, pickled radish and carrot until fragrant. Mix in the seasoning. Give a good stir-fry. Sprinkle with the spring onion. Set aside.
4. Dust both hands with caltrop starch. Knead the mud carp meat mixture into flat and round shape. Wrap the BBQ pork stuffing. Fold into triangles.
5. Bring the seasoned stock to the boil. Put in the fish dumplings. Bring to the boil. Add the Chinese kale and cook for a while. Sprinkle with the yellow chives. Serve.

扣三絲
Steamed Ham, Egg and Black Mushroom Shreds

小秘訣 Tips

豆乾絲於南貨店有售，或可用粉絲代替。

Dried tofu is available at the shop selling Shanghainese food, or you may use mung bean vermicelli instea

材料
Ingredients

火腿片 4 兩
雞蛋 3 個
大冬菇 2 兩（處理方法參考 p.12）
豆乾絲 2 兩
葱 1 條
薑 2 片
菜心 4 兩

150 g sliced ham
3 eggs
75 g large dried black mushrooms
(refer to p.12 for method)
75 g shredded dried tofu
1 sprig spring onion
2 slices ginger
150 g flowering Chinese cabbage

調味料
Seasoning

鹽半茶匙
糖 1/4 茶匙
上湯 1 安士（約 28 毫升）
紹酒 1 茶匙
麻油少許

1/2 tsp salt
1/4 tsp sugar
1 oz. (28 ml) stock
1 tsp Shaoxing wine
sesame oil

獻汁
Thickening glaze

上湯 3 安士（約 85 毫升）
生粉半茶匙

3 oz. (85 ml) stock
1/2 tsp caltrop starch

做法
Method

1. 雞蛋拂勻，下少許鹽及糖略拌，煎成薄蛋皮，切成長幼絲。
2. 火腿切成長幼絲；預留一隻冬菇，其餘切成幼絲。
3. 燒滾水，放入豆乾絲飛水，瀝乾水分。
4. 在深碗內均勻地塗上油，碗底放入冬菇一隻，碗邊分別排上蛋絲、火腿絲及冬菇絲，中間位置壓入豆乾絲，按緊。
5. 調味料拌勻，傾入步驟 4 的碗內，薑葱鋪面蒸約半小時。
6. 將蒸熟的材料反扣碟上，排上已灼熟之菜心，倒入煮滾的獻汁即可。

1. Whisk the eggs. Mix with salt and sugar. Fry into thin sheet. Finely cut into long shreds.
2. Finely cut the ham into long shreds. Keep a black mushroom and finely shred the rest.
3. Bring water to the boil. Scald the dried tofu shreds and drain.
4. Spread oil inside a deep bowl evenly. Put a black mushroom on the bottom. Place the shredded egg, ham and black mushroom separately along the edge. Put the dried tofu shreds in the middle. Press firmly.
5. Mix the seasoning evenly. Pour into the bowl from Step 4. Lay the ginger and spring onion on top. Steam for about 1/2 hour.
6. Reverse the bowl on a plate. Remove the bowl. Leave the steamed ingredients on the plate. Put the flowering Chinese cabbage in the side of the plate. Pour in the boiled thickening glaze. Serve.

四川陳皮雞

Stir-fried Spicy Chicken in Sichuan Style

小秘訣Tips

- 這款四川菜式比較辣，可隨意加減辣椒份量。
- 用滾油爆香花椒後，熱油帶陣陣花椒香氣，可棄去花椒粒，香氣已足夠。
- This Sichuan cuisine is rather spicy. You may adjust the amount of chillies to suit your taste.
- The hot oil, which has been used to stir-fry the Sichuan peppercorns, smells the aroma of Sichuan peppercorn. As it is fragrant enough, they can be dumped.

材料 Ingredients

雞半隻
紅辣椒乾 3 隻
陳皮半個（切絲）
花椒 15 粒
乾葱肉 10 粒
薑 3 片
葱 3 條（切段）

1/2 chicken
3 dried red chillies
1/2 dried tangerine peel
(shredded)
15 Sichuan peppercorns
10 shallots
3 slices ginger
3 sprigs spring onion
 (cut into sections)

醃料 Marinade

老抽半湯匙
鹽半茶匙
胡椒粉少許
生粉 1 湯匙（後下）

1/2 tbsp dark soy sauce
1/2 tsp salt
ground white pepper
1 tbsp caltrop starch (added last)

調味料 Seasoning

糖 2 茶匙
老抽 1/4 茶匙
生抽、麻油及鹽各少許
水半杯

2 tsps sugar
1/4 tsp dark soy sauce
light soy sauce
sesame oil
salt
1/2 cup water

做法 Method

1. 雞洗淨，斬件，下醃料拌勻醃 15 分鐘，下生粉略拌，放入滾油內炸至稍乾，盛起，瀝乾油分。

2. 陳皮絲、辣椒乾、薑片及乾葱肉放入滾油炸至稍乾，盛起備用。

3. 燒熱油 1 湯匙，下花椒爆香，棄去，加入調味料、雞件及步驟 2 配料炒勻，最後下葱段拌勻，即可享用。

1. Wash the chicken. Chop into pieces. Mix with the marinade and rest for 15 minutes. Roughly mix with the caltrop starch. Deep-fry in scorching oil until the chicken is a bit dry. Drain.

2. Deep-fry the dried tangerine peel, dried red chillies, ginger and shallots in scorching oil until they are a bit dry. Set aside.

3. Heat 1 tbsp of oil. Stir-fry the Sichuan peppercorns until aromatic. Dump the Sichuan peppercorns. Add the seasoning, chicken and the spices from Step 2. Give a good stir-fry. Finally mix in the spring onion. Serve.

百花龍鳳卷
Deep-fried Minced Shrimp, Pork and Chicken Liver Rolls

材料 Ingredients

蝦肉 6 兩（蝦膠處理方法參考 p.14）
肥豬肉少許（煮熟、切粒）
燒雞肝 3 兩
方包 4 片

225 g shelled shrimps (refer to p.14 for method)
a little fat pork (cooked; diced)
113 g roast chicken liver
4 pieces sandwich bread

調味料 Seasoning

生粉 1 1/2 湯匙
蛋白半個
鹽半茶匙
胡椒粉及麻油各少許

1 1/2 tbsps caltrop starch
1/2 egg white
1/2 tsp salt
ground white pepper
sesame oil

做法 Method

1. 蝦膠、肥肉粒及調味料順一方向拌起膠，冷藏半小時。
2. 燒雞肝切成長條狀，備用。
3. 在方包上均勻地抹上蝦膠，放入燒肝捲成長卷形，下油鑊用中火炸至黃色，切件享用。

1. Mix the minced shrimp, fat pork a[nd] seasoning together. Stir in one directi[on] until gluey. Chill for 1/2 hour.
2. Cut the roast chicken liver into stri[ps.] Set aside.
3. Spread the shrimp mixture evenly on [the] sandwich bread. Put in the roast chick[en] liver. Roll into a cylinder. Deep-fry in [oil] over medium heat until golden. Cut i[nto] pieces and serve.

小秘訣 Tips

燒雞肝於燒臘店有售；若買不到的話，可將鮮[肝]洗淨，用叉燒醬及紹酒等醃味，放入焗爐烤熟[，效]果相同。

Roast chicken liver can be bought at roast [meat] shops. If not available, you can use the fresh [liver.] Wash the chicken liver, marinate with barb[ecue] sauce, Shaoxing wine and so on, and then ba[ke in] an oven. It has the same effect.

紅燒魚唇
Braised Fish Snout

材料
Ingredients

急凍魚唇 12 兩
冬菇 8 隻（處理方法參考 p.12）
筍肉 4 兩
薑 4 片
蔥 2 條（切段）

450 g frozen fish snout
8 dried black mushrooms (refer to p.12 for method)
150 g skinned bamboo shoot
4 slices ginger
2 sprigs spring onion (cut into sections)

調味料
Seasoning

鹽及糖各半茶匙
蠔油 2 湯匙
生抽 1 茶匙
胡椒粉及麻油各少許
水 1 1/2 杯

1/2 tsp salt
1/2 tsp sugar
2 tbsps oyster sauce
1 tsp light soy sauce
ground white pepper
sesame oil
1 1/2 cups water

獻汁
Thickening glaze

生粉 1 茶匙
老抽 1 茶匙
水 1 湯匙
* 拌勻

1 tsp caltrop starch
1 tsp dark soy sauce
1 tbsp water
* mixed well

做法
Method

1. 筍肉飛水，用凍開水浸涼，切成厚件；冬菇切半，備用。

2. 魚唇解凍，洗淨，放入盛有 3 杯熱水的煲內，下薑 2 片、蔥 2 條及紹酒 1 湯匙煮滾約 15 分鐘，盛起，洗淨，切件。

3. 燒熱油 3 湯匙，爆香薑片、筍肉及冬菇，灒酒，下調味料及魚唇煮約 15 分鐘，至汁液濃稠，埋獻煮滾，最後加入蔥段及麻油拌勻即可。

1. Scald the bamboo shoot. Soak in cold drinking water to cool. Cut into thick pieces. Cut the black mushrooms in half. Set aside.

2. Defrost the fish snout and rinse. Cook in 3 cups of hot water with 2 slices of ginger, 2 sprigs of spring onion and 1 tbsp of Shaoxing wine for about 15 minutes. Rinse and cut into pieces.

3. Heat 3 tbsps of oil. Stir-fry the ginger, bamboo shoot and black mushrooms until sweet-scented. Sprinkle with wine. Add the seasoning and fish snout. Cook for about 15 minutes, or until the sauce is thick. Mix in the thickening glaze. Bring to the boil. Finally sprinkle with the spring onion and sesame oil. Mix up. Serve.

鉢頭膏蟹飯

Steamed Crab with Glutinous Rice in Earthen Bowl

材料 / Ingredients

膏蟹（大）1 隻
蝦仁 2 兩
糯米 8 兩
冬菇 6 隻
（處理方法參考 p.12）
冬筍 2 兩
葱 2 條

1 large female mud crab
75 g shelled shrimps
300 g glutinous rice
6 dried black mushrooms
(refer to p.12 for method)
75 g winter bamboo shoots
2 sprigs spring onion

調味料 / Seasoning

鹽及糖各半茶匙
生抽半茶匙
胡椒粉及麻油各少許

1/2 tsp salt
1/2 tsp sugar
1/2 tsp light soy sauce
ground white pepper
sesame oil

醬油料 / Sauce

生抽 2 茶匙
老抽及糖各 1 茶匙
麻油少許
水 2 湯匙

2 tsps light soy sauce
1 tsp dark soy sauce
1 tsp sugar
sesame oil
2 tbsps water

做法 / Method

1. 糯米洗淨，用水浸 2 小時，放入已抹油的瓦鉢，下適量水蒸 20 分鐘至熟 (水蓋過米粒)。
2. 蝦仁、冬菇及冬筍切碎，用油炒熟，加入調味料拌炒，灒酒，盛起，與糯米飯拌勻，備用。
3. 膏蟹洗淨，斬件，放在糯米飯上蒸約 10 分鐘，灑入葱粒，澆上用油煮熱之醬油即可。

1. Rinse the glutinous rice. Soak in water for 2 hours. Put into a greased earthen bowl. Put in some water to cover the rice. Steam for 20 minutes, or until done.

2. Finely chop the shrimps, black mushrooms and winter bamboo shoots. Stir-fry with oil until done. Add the seasoning and stir-fry. Sprinkle with wine. Mix with the glutinous rice. Set aside.

3. Wash the female mud crab. Chop into pieces. Put on top of the glutinous rice. Steam for about 10 minutes. Sprinkle with the diced spring onion. Pour the hot sauce (cooked with oil) on top. Serve.

小秘訣 Tips

- 糯米黏糯，胃痛者不宜多吃，可改用粘米。
- 澆上醬油後，可將瓦鉢放在明火或焗爐烘香。
- 沒有瓦鉢的話，可用瓦煲代替。
- Glutinous rice is quite sticky, and so it is not so suitable for people having stomachache. Regular rice can be used instead.
- After drizzling with the cooked sauce, you may put the earthen bowl on the flame to toast until fragrant, or bake it in on oven.
- If you have no earthen bowl, use a casserole instead.

銀杏蓮子雞
Braised Chicken with Gingkoes and Lotus Seeds

材料 Ingredients

雞半隻
鮮蓮子及鮮銀杏各 2 兩
花旗參片 5 錢
杞子 2 湯匙
薑 3 片

1/2 chicken
75 g fresh lotus seeds
75 g fresh gingkoes
19 g sliced American ginseng
2 tbsps Qi Zi
3 slices ginger

調味料 Seasoning

糖半茶匙
鹽適量

1/2 tsp sugar
salt

小秘訣 Tips

杞子不可久煮，否則帶酸味，最後加入略煮即可。

Qi Zi will give a sour taste if cooked for a long time. Just cook it in the final stage.

做法 Method

1. 雞洗淨，斬件，放入滾水內飛水，用水沖淨，備用。
2. 蓮子、銀杏及花旗參洗淨；杞子略洗。
3. 燒紅鑊，下少許油及薑片，爆香雞件，加入蓮子、銀杏及花旗參，潷酒，傾入水 3 杯及調味料，以小火煮約 20 分鐘，最後加入杞子略煮 5 分鐘即可。

1. Wash the chicken. Chop into pieces. Blanch and then rinse. Set aside.
2. Rinse the lotus seeds, gingkoes and American ginseng. Slightly rinse Qi Zi.
3. Heat a wok. Put in a little oil and the ginger. Stir-fry the chicken until fragrant. Add the lotus seeds, gingkoes and American ginseng. Sprinkle with wine. Pour in 3 cups of water and the seasoning. Simmer for about 20 minutes. Finally add Qi Zi. Cook for 5 minutes. Serve.

燴牛脷
Braised Beef Tongue

材料
Ingredients

牛脷 1 條（約 2 斤）
薑粒、蒜茸及葱粒各 1 湯匙
1 beef tongue (about 1.2 kg)
1 tbsp diced ginger
1 tbsp finely chopped garlic
1 tbsp diced spring onion

小秘訣 Tips

- 牛脷用白醋水煮後，即浸入冰水內，容易刮出表面不潔之物。
- 洋葱及甘筍必須炒香，湯汁才香濃美味。
- Soak the beef tongue in ice water immediately after cooking it with white vinegar. The dirt on its surface can be scraped off easily.
- The onion and carrot must be stir-fried until fragrant. The soup stewed will have a rich smell and taste. It is sweet-scented and delicious.

上湯料
Ingredient for stock

大洋葱 1 個（切片）
大甘筍 1 個（切件）
香葉 3 片
八角 2 粒
黑椒粒 1 湯匙
冰糖 1 湯匙
白酒 3 湯匙
鹽 6 茶匙
清雞湯 16 杯

1 large onion (sliced)
1 large carrot (cut into pieces)
3 bay leaves
2 star anises
1 tbsp black peppercorns
1 tbsp rock sugar
3 tbsps white wine
6 tsps salt
16 cups chicken stock

調味汁
Seasoning

白米醋 1 茶匙
粟粉 2 茶匙
生抽及白酒各 1 茶匙
燜牛脷湯 1 杯
麻油 1 茶匙
胡椒粉少許

1 tsp white rice vinegar
2 tsps corn flour
1 tsp light soy sauce
1 tsp white wine
1 cup beef tongue cooking soup
1 tsp sesame oil
ground white pepper

做法
Method

1. 牛脷用水浸 1 小時，下鹽 1 湯匙擦洗，放入加有 1 湯匙白醋的熱水內煮 20 分鐘，再浸冰水至涼，刮去牛脷表面之不潔。

2. 燒熱油，下洋葱及甘筍炒香，潷白酒，加入其餘上湯料浸過牛脷，用慢火煮約 1 1/2 小時，關火，加蓋浸 1 小時，取出切片上碟，燜牛脷湯留用。

3. 燒熱少許油，下蒜茸、葱粒及薑粒炒香，加入白酒及燜牛脷湯煮滾，下其餘調味汁料略滾，澆在牛脷片上即成。

1. Soak the beef tongue in water for 1 hour. Rub with 1 tbsp of salt. Cook in hot water with 1 tbsp of white vinegar for 20 minutes. Soak in ice water to cool. Scrape off the dirt on the surface of the beef tongue.

2. Heat oil. Stir-fry the onion and carrot until fragrant. Sprinkle with the white wine. Put in the remaining stock ingredients. The liquid should be enough to cover the beef tongue. Simmer for about 1 1/2 hours. Turn off heat. Soak for 1 hour with a lid on. Cut into slices and arrange on the plate. Keep the cooking soup for later use.

3. Heat a little oil. Stir-fry the garlic, spring onion and ginger until aromatic. Add the white wine and the beef tongue cooking soup. Bring to the boil. Put in the remaining seasoning and slightly boil. Pour on the beef tongue. Serve.

發財好市
Stewed Dried Oysters with Black Moss

材料 Ingredients

蠔豉 3 兩
冬菇 1 兩（處理方法參考
p.12）
髮菜半兩
生菜半斤
薑 2 片
蒜茸 1 茶匙

113 g dried oysters
38 g dried black mushrooms
(refer to p.12 for method)
19 g dried black moss
300 g lettuce
2 slices ginger
1 tsp finely chopped garlic

煨料 Ingredients for stewing

薑 2 片
葱 1 條
紹酒適量

2 slices ginger
1 sprig spring onion
Shaoxing wine

調味料 Seasoning

上湯 1 杯
蠔油 2 湯匙
糖及老抽各 1 茶匙
鹽 1/4 茶匙
胡椒粉及麻油各少許

1 cup stock
2 tbsps oyster sauce
1 tsp sugar
1 tsp dark soy sauce
1/4 tsp salt
ground white pepper
sesame oil

獻汁 Thickening glaze

生粉 1 茶匙
水 2 湯匙
* 拌勻

1 tsp caltrop starch
2 tbsps water
* mixed well

小秘訣 Tips

蠔豉以蠔身飽滿，色澤金黃油潤為佳。
Choose dried oysters which are plump, golden and glossy.

做法
Method

1. 蠔豉洗淨，用水略浸，放入滾水內，下煨料煨煮片刻，盛起，水留用。

2. 髮菜用水浸 10 分鐘，洗淨，擠乾水分，放入上述煨料水內煮 3 分鐘，盛起，瀝乾水分。

3. 燒熱油 3 湯匙，爆香薑片及蒜茸，下蠔豉、冬菇及髮菜炒勻，潷酒，加入調味料用小火燜 15 分鐘，埋獻，鋪在已灼的生菜面即可上桌。

1. Rinse the dried oysters. Soak in water for a while. Put into boiling water. Add the ingredients for stewing and cook for a moment. Dish up. Keep the cooking water.

2. Soak the dried black moss in water for 10 minutes. Rinse well. Squeeze water out. Put into the above cooking water. Cook for 3 minutes. Drain.

3. Heat 3 tbsps of oil. Stir-fry the ginger and garlic until aromatic. Add the dried oysters, black mushrooms and black moss. Give a good stir-fry. Sprinkle with wine. Put in the seasoning. Simmer for 15 minutes. Mix in the thickening glaze. Lay on top of the blanched lettuce. Serve.

翠玉手撕雞

Shredded Chicken with Jellyfish and Vegetables

小秘訣 Tips

炮製滑嫩的雞肉，秘訣在於煲滾後再加蓋焗透。

To make silky chicken meat, the secret is to bring the liquid with chicken to the boil, and then leave it with a lid on to let the heat make it done.

材料
Ingredients

雞 1 隻
海蜇 1 斤
西芹 1/4 棵
青瓜及甘筍各半條
白芝麻 2 湯匙（白鑊烘香）

1 chicken
600 g jellyfish
1/4 stalk celery
1/2 cucumber
1/2 carrot
2 tbsps white sesame seeds
(toasted in dry wok)

滷水料
Chinese marinade

鹽及生抽各 2 湯匙
糖 1 湯匙
薑 3 片
葱 2 條
八角 2 粒
紹酒 3 湯匙
水 8 杯

2 tbsps salt
2 tbsps light soy sauce
1 tbsp sugar
3 slices ginger
2 sprigs spring onion
2 star anises
3 tbsps Shaoxing wine
8 cups water

醬汁
Sauce

甜酸雞醬適量
芝麻醬 3 湯匙

sweet and sour sauce for chicken
3 tbsps sesame sauce

做法
Method

1. 滷水料煮滾，調至小火，煲約 10 分鐘。
2. 雞洗淨，抹乾水分，放入滷水汁內煮滾，關火，浸約 20 分鐘，待涼，撕出雞肉。
3. 海蜇用水浸洗，切件，放入滾水內灼 2 分鐘至捲曲，取出，用冷開水浸至涼，瀝乾水分。
4. 青瓜、甘筍及西芹洗淨，抹乾水分，切絲鋪在碟內，排上雞絲及海蜇，澆上麻油、甜酸雞醬及芝麻醬，灑上烘香白芝麻即可。

1. Bring the Chinese marinade to the boil. Turn down the heat. Simmer for about 10 minutes.
2. Wash the chicken and wipe dry. Put into the Chinese marinade. Bring to the boil. Turn off heat. Soak for about 20 minutes. Let it cool down. Tear off the chicken meat.
3. Soak the jellyfish in water and wash. Cut into pieces. Blanch for 2 minutes, or until it curls up. Take out. Soak in cold drinking water to cool. Drain well.
4. Rinse the cucumber, carrot and celery. Wipe dry. Cut into shreds. Lay on a plate. Top with the chicken and jellyfish. Pour the sesame oil, sweet and sour sauce for chicken and sesame sauce on top. Sprinkle with the toasted sesame seeds. Serve.

清蒸荷葉魚
Steamed Grouper on Lotus Leaf

材料 Ingredients

石斑肉 1 斤
冬菇（大）1 兩
（處理方法參考 p.12）
金華火腿 3 兩
荷葉 1 張
薑絲 2 湯匙

600 g grouper meat
38 g large dried black mushrooms
 (refer to p.12 for method)
113 g Jinhua ham
1 lotus leaf
2 tbsps shredded ginger

醃料 Marinade

鹽 1 茶匙
糖半茶匙
生粉 1 1/2 茶匙
胡椒粉及麻油各少許

1 tsp salt
1/2 tsp sugar
1 1/2 tsps caltrop starch
ground white pepper
sesame oil

做法 Method

1. 荷葉放入滾水內略灼，洗淨，抹乾水分。
2. 金華火腿用滾水煮片刻，切片，加入糖及紹酒各 1 茶匙，隔水蒸 15 分鐘。
3. 石斑肉洗淨，抹乾水分，切件，下醃料拌勻，備用。
4. 碟內鋪上荷葉，依序排上金華火腿、魚肉、冬菇及薑絲，隔水蒸 10 分鐘，上桌品嘗。

1. Slightly blanch the lotus leaf. Rinse and wipe dry.
2. Cook the Jinhua ham in boiling water for a while. Cut into slices. Mix with 1 tsp each of sugar and Shaoxing wine. Steam for 15 minutes.
3. Wash the grouper meat. Wipe dry. Cut into pieces. Mix with the marinade. Set aside.
4. Lay the lotus leaf on a plate. Arrange the Jinhua ham, fish, black mushrooms and shredded ginger on the leaf in sequence. Steam for 10 minutes. Serve.

小秘訣 Tips

乾荷葉比新鮮荷葉清香，而且方便購買，雜貨店有售。

Dried lotus leaf has a lighter aroma than fresh lotus leaf. It is also easily available at the groceries.

冬瓜釀帶子
Winter Melon Stuffed with Scallops

材料
Ingredients

冬瓜 1 1/2 斤
（圓環狀，約 1 吋高）
急凍帶子 10 粒
髮菜半兩（用水浸透）
甘筍片適量

900 g winter melon (in ring
shape; about 1-inch high)
10 frozen scallops
19 g dried black moss
(soaked in water to soften)
carrot slices

煨料
Ingredients for stewing

油 1 湯匙
紹酒、薑、葱各適量

1 tbsp oil
Shaoxing wine
ginger
spring onion

獻汁
Thickening glaze

上湯半杯
鹽及糖各 1/4 茶匙
胡椒粉及麻油各少許
生粉 1 1/2 茶匙

1/2 cup stock
1/4 tsp salt
1/4 tsp sugar
ground white pepper
sesame oil
1 1/2 tsps caltrop starch

做法
Method

1. 燒滾水 2 杯，下髮菜及煨料煮片刻，瀝乾水分。
2. 冬瓜去籽、洗淨，抹上鹽半茶匙，用圓型模在瓜面相間地刮出瓜肉，放在碟內隔水蒸半小時。
3. 髮菜用少許油及鹽拌勻，隔水蒸半小時，備用。
4. 帶子解凍，抹乾水分，加入少許鹽、生粉及胡椒粉拌勻，釀入瓜面小孔內，相間地鋪上甘筍片，中央位置放上髮菜，蒸 10 分鐘。
5. 燒熱油 2 湯匙，下獻汁煮滾，澆在冬瓜面即可。

1. Bring 2 cups of water to the boil. Put in the black moss and ingredients for stewing. Cook for a while. Drain.
2. Remove the seeds of the winter melon and rinse. Spread with 1/2 tsp of salt. Scrape out the flesh alternately with a round mold. Put the winter melon on a plate. Steam for 1/2 hour.
3. Mix the black moss with a little oil and salt. Steam for 1/2 hour. Set aside.
4. Defrost the scallops. Wipe dry. Mix with a little salt, caltrop starch and ground white pepper. Stuff into the small holes on the surface of the winter melon. Lay the carrot slices on the winter melon alternately. Put the black moss in the middle. Steam for 10 minutes.
5. Heat 2 tbsps of oil. Put in the thickening glaze. Bring to the boil. Pour over the winter melon. Serve.

小秘訣 Tips

瓜面上小孔的大小應與帶子相同，釀入帶子後才美觀好看。

The winter melon stuffed with scallops will look gorgeous by making the small holes in the same size as that of the scallops.

香橙芝麻牛仔骨

Fried Beef Short Ribs in Orange Sauce with Sesame Seeds

材料
Ingredients

牛仔骨 450 克
白芝麻 1 湯匙（烘香）
橙皮絲 2 湯匙（熱水略泡）

450 g beef short ribs
1 tbsp white sesame seeds
(toasted)
2 tbsps shredded orange zest
(slightly soaked in hot water)

醃料
Marinade

油及糖各 1 茶匙
生抽 1 1/2 湯匙
香橙酒或紹酒少許
胡椒粉少許

1 tsp oil
1 tsp sugar
1 1/2 tbsps light soy sauce
orange liqueur or Shaoxing wine
ground white pepper

調味料
Seasoning

香橙酒 1 湯匙
橙汁 3 湯匙
橙皮果醬 2 茶匙
糖 1 茶匙
生抽半茶匙

1 tbsp orange liqueur
3 tbsps orange juice
2 tsps orange marmalade
1 tsp sugar
1/2 tsp light soy sauce

做法
Method

1. 牛仔骨解凍，洗淨，抹乾水分，下醃料拌勻醃 15 分鐘，撲上粟粉，放入熱油內煎至金黃色，備用。

2. 燒熱半湯匙油，下調味料用小火煮勻，加入橙皮絲煮至汁液濃稠，放下牛仔骨炒勻，最後灑上已烘香白芝麻即成。

1. Defrost the beef short ribs. Rinse well and wipe dry. Mix with the marinade and rest for 1 minutes. Coat with corn flour. Fry in hot oil until golden. Set aside.

2. Heat 1/2 tbsp of oil. Put in the seasoning. Cook over low heat until it mixes up. Add the shredded orange zest. Cook until the sauce is thick. Put in the beef short ribs. Give a good stir fry. Finally sprinkle with the toasted white sesame seeds. Serve.

小秘訣 Tips

牛仔骨容易煎熟，轉成金黃色即上碟，否則肉質變韌。

It is easy for the beef short ribs to be cooked through. When they turn golden, remove at once. The meat texture will be tough if cooked for a long time.

越南香茅椰汁雞

Vietnamese Stewed Chicken with Lemongrass in Coconut Sauce

小秘訣 Tips

馬鈴薯泡油炸至金黃色，鎖住外層，燜煮時可保持薯塊完整。

The potato will be coated by deep-frying in oil until golden. It can keep the potato intact while stewing.

材料
Ingredients

鮮雞半隻
香茅2棵（切段）
馬鈴薯2個（切件）
洋葱1個（切塊）
乾葱2粒（剁茸）
蒜頭1粒（剁茸）
黃薑粉及南薑粉各1茶匙
椰汁2湯匙
椰汁3安士（約85毫升）
花奶4安士（約115毫升）
上湯2杯

1/2 chicken
2 stalks lemongrass (cut into sections)
2 potatoes (cut into pieces)
1 onion (cut into pieces)
2 shallots (finely chopped)
1 clove garlic (finely chopped)
1 tsp ground turmeric
1 tsp ground galangal
2 tbsps coconut milk
3 oz. (85 ml) coconut milk
4 oz. (115 ml) evaporated milk
2 cups stock

醃料
Marinade

鹽1茶匙
糖半茶匙
生粉1茶匙

1 tsp salt
1/2 tsp sugar
1 tsp caltrop starch

做法
Method

1. 雞洗淨，斬件，下醃料拌勻醃半小時。
2. 馬鈴薯放入熱油內略泡油至金黃色，盛起。
3. 燒熱鑊，下油2湯匙，加入洋葱及鹽半茶匙爆香，放入香茅、黃薑粉、南薑粉、乾葱茸及蒜茸炒香，下雞件不斷炒勻。
4. 傾入椰汁2湯匙拌炒，再下上湯1杯及馬鈴薯炒勻，放入餘下上湯用大火燜15分鐘，蓋不要緊閉，最後放入花奶及椰汁煮滾即可上碟。

1. Wash the chicken. Chop into pieces. Mix with the marinade and rest for 1/2 hour.
2. Deep-fry the potato in hot oil slightly until golden. Set aside.
3. Heat a wok. Put in 2 tbsps of oil. Stir-fry the onion with 1/2 tsp of salt until fragrant. Add the lemongrass, ground turmeric, ground galangal, shallots and garlic. Stir-fry until aromatic. Put in the chicken and keep stir-frying until the ingredients are mixed up.
4. Pour in 2 tbsps of the coconut milk. Stir-fry for a while. Add 1 cup of the stock and the potato. Give a good stir-fry. Put in the rest stock. Cook over high heat for 15 minutes. Leave a slit between the wok and the lid. Finally put in the evaporated milk and coconut milk. Bring to the boil. Serve.

百花釀鳳翼
Stuffed Chicken Wings with Minced Shrimp

材料 Ingredients

雞全翼 6 隻
蝦肉半斤
（蝦膠處理方法參考 p.14）
肥豬肉 1 湯匙
雞蛋 1 個
6 whole chicken wings
300 g shelled shrimps
(refer to p.14 for method)
1 tbsp fat pork
1 egg

醃料 Marinade

鹽半茶匙
胡椒粉少許
1/2 tsp salt
ground white pepper

調味料 Seasoning

鹽半茶匙
生粉及蛋白各 1 湯匙
胡椒粉及麻油各少許
1/2 tsp salt
1 tbsp caltrop starch
1 tbsp egg white
ground white pepper
sesame oil

做法 Method

1. 肥肉用水灼熟，切成幼粒。
2. 蝦膠、肥肉粒及調味料順一方向拌勻，撻至起膠及有黏性，冷藏片刻。
3. 雞翼用鹽水浸洗，抹乾水分，去掉雞骨，拌醃料略醃。
4. 將蝦膠釀入雞翼髀內，撲上生粉，下油鍋內炸至金黃色，瀝乾油分，即可品嘗。

1. Blanch the fat pork until done. Finely dice.
2. Stir the minced shrimp, fat pork and seasoning evenly in one direction. Throw into a bowl repeatedly until sticky. Chill for a moment.
3. Soak the chicken wings in salted water and wash. Wipe dry. Bone the chicken wings. Mix with the marinade. Leave for a while.
4. Stuff the minced shrimp mixture into the upper part of the chicken wing. Coat with caltrop starch. Deep-fry in oil until golden. Drain. Serve hot.

小秘訣 Tips

- 雞全翼去骨的方法，與雞中翼方法相同，用小刀慢慢褪出雞肉，折斷關節即可。若沒把握的話，可改用雞中翼。

- 別將蝦膠釀得太漲滿，約七成即可，因雞翼熟透後，肉質會收縮。

- The method of boning chicken's whole wings is the same as boning mid-joint wings. Remove the bone from the meat slowly with a small knife, and then break the joints. If you have no confidence in doing so, use the chicken mid-joint wings instead.

- Do not stuff too much minced shrimp mixture into the chicken wings. Just make it 70% full. The meat will shrink after the chicken wings are fully cooked.

金腿伴雙花
Jinhua Ham with Broccoli and Cauliflower

材料
Ingredients

西蘭花 1 個
椰菜花半個
金華火腿 2 兩
上湯 2 杯

1 broccoli
1/2 cauliflower
75 g Jinhua ham
2 cups stock

調味料
Seasoning

鹽 1/3 茶匙
糖 1/4 茶匙
胡椒粉及麻油各少許
上湯 2 安士（約 57 毫升）

1/3 tsp salt
1/4 tsp sugar
ground white pepper
sesame oil
2 oz. (57 ml) stock

獻汁
Thickening glaze

生粉 1 茶匙
水 2 湯匙
* 拌勻

1 tsp caltrop starch
2 tbsps water
* mixed well

做法
Method

1. 西蘭花及椰菜花切成小棵，用鹽水略浸，洗淨，放入滾水內，加糖、紹酒、鹽及油略灼熟，盛起，瀝乾水分。

2. 燒滾上湯，下西蘭花及椰菜花煨煮至腍，盛起，排於碟上。

3. 金華火腿用水煮約 10 分鐘，洗淨，拌入糖及紹酒各少許，蒸約 15 分鐘，切薄片，相間排於雙花中。

4. 燒熱油，下調味料煮滾，埋獻，澆在菜面即成。

1. Cut the broccoli and cauliflower into small floral pieces. Soak in salted water for a while and rinse. Put into boiling water. Add sugar, Shaoxing wine, salt and oil. Slightly blanch and drain.

2. Bring the stock to the boil. Cook the broccoli and cauliflower until tender. Arrange alternately on a plate.

3. Cook the Jinhua ham in water for about 10 minutes and rinse. Mix with a little sugar and Shaoxing wine. Steam for about 15 minutes. Finely slice. Arrange between every row of broccoli and cauliflower.

4. Heat oil. Add the seasoning. Bring to the boil. Mix in the thickening glaze. Pour over the vegetables. Serve.

皇帝蝦
Deep-fried Prawns with Egg Yolks

材料
Ingredients

中蝦 1 斤
鹹蛋 6 個（只取蛋黃）
雞蛋白 2 個（拂勻）
雞蛋黃 3 個
生粉 2 湯匙

600 g medium prawns
6 salted eggs
(the egg yolks only)
2 egg whites (whisked)
3 egg yolks
2 tbsps caltrop starch

醃料
Marinade

鹽半茶匙
胡椒粉及麻油各少許

1/2 tsp salt
ground white pepper
sesame oil

做法
Method

1. 鹹蛋黃蒸熟，搓爛備用。
2. 蝦洗淨，抹乾水分，下醃料拌勻醃半小時。
3. 蝦依序蘸上生粉及蛋白，下油鍋炸至金黃色。
4. 燒熱油，下鹹蛋黃茸用慢火不斷推炒至起泡，關火，加入已炸脆之蝦拌勻。
5. 雞蛋黃盛於不鏽鋼盆內，隔水蒸熟，不停攪拌至呈蟹膏狀，鋪在蝦面上即成。

1. Steam the salted egg yolks until done. Mash up. Set aside.
2. Rinse the prawns and wipe dry. Mix with the marinade and rest for 1/2 hour.
3. Dip the prawns into the caltrop starch and then the egg whites. Deep-fry in oil until golden. Set aside.
4. Heat oil. Put in the mashed salted egg yolk. Keep pushing and stir-frying over low heat until it bubbles. Turn off heat. Add the prawns and mix up.
5. Place the egg yolks in a stainless steel dish. Steam until done. Keep stirring until it looks like crab roe. Lay on top of the prawns. Serve.

小秘訣 Tips

炒鹹蛋黃必須用油不斷推炒至起泡，至散發香氣即可。

When stir-frying the mashed salted egg yolk with oil, you must keep pushing and stir-frying until it bubbles and smells fantastic.

香芋葡汁鴨

Braised Duck and Taro in
Curry-coconut Sauce

小秘訣 Tips

最後加入椰汁及花奶，略煮即上碟，口感香滑。

Slightly cook the coconut milk and evaporated milk at the final stage. It is fragrant and creamy.

材料
Ingredients

鴨半隻
荔甫芋半斤
洋蔥 1 個
咖喱醬 3 湯匙
花奶 3 安士（約 85 毫升）
椰汁 6 安士（約 170 毫升）

1/2 duck
300 g taro
1 onion
3 tbsps curry sauce
3 oz. (85 ml) evaporated milk
6 oz. (170 ml) coconut milk

醃料
Marinade

鹽及糖各半茶匙
生粉、生抽、薑汁及紹酒各 1 湯匙

1/2 tsp salt
1/2 tsp sugar
1 tbsp caltrop starch
1 tbsp light soy sauce
1 tbsp ginger juice
1 tbsp Shaoxing wine

做法
Method

1. 鴨洗淨，斬件，拌入醃料待片刻。
2. 芋頭去皮，洗淨，切件，略泡油至金黃色，盛起備用。
3. 洋蔥撕去外衣，切成塊。
4. 燒熱油，爆香洋蔥塊，加入咖喱醬炒香，下鴨件不斷炒勻，轉放瓦煲內，傾入水蓋過鴨件，中火燜 15 分鐘，放入芋頭再燜 20 分鐘，最後加入椰汁及花奶煮滾即可。

1. Wash the duck. Chop into pieces. Mix with the marinade and rest for a while.
2. Peel and rinse the taro. Cut into pieces. Slightly deep-fry in oil until golden. Set aside.
3. Tear the skin off the onion. Cut into pieces.
4. Heat oil. Stir-fry the onion until fragrant. Add the curry sauce and stir-fry until sweet-scented. Put in the duck and keep stir-frying until the ingredients are mixed up. Transfer to a casserole. Pour in water to cover the duck. Cook over medium heat for 15 minutes. Put in the taro and cook again for 20 minutes. Finally add the coconut milk and evaporated milk. Bring to the boil. Serve.

羅漢上素

Stir-fried Mixed Vegetables and Mushrooms

材料 Ingredients

鮮冬菇 10 隻
雲耳 10 朵
鮑魚菇 8 個
草菇 10 粒
蘑菇 10 粒
小粟米 8 支
菜心 4 兩
甘筍（小）1 個
炸生筋 10 個

10 fresh black mushrooms
10 cloud ear fungus
8 oyster mushrooms
10 straw mushrooms
10 button mushrooms
8 baby corn
150 g flowering Chinese cabbage
1 small carrot
10 pieces deep-fried gluten balls

小秘訣 Tips

• 市面有各種罐裝菇類，但建議購買鮮菇烹調，爽口、營養又豐富。

• 先將菇類飛水，可去除草腥味之餘，也縮短回鑊的炒煮時間。

• There is a collection of canned mushrooms in the market. Using fresh mushrooms to make this dish is recommended. They are crunchy and very nutritious.

• Scald the mushrooms beforehand to remove their grassy smell. It can also shorten the time for stir-frying.

調味料
Seasoning

獻汁
Thickening glaze

齋蠔油 2 湯匙
老抽 1 湯匙
鹽半茶匙
糖 1 茶匙
上湯 2 湯匙
麻油半茶匙

2 tbsps vegetarian oyster sauce
1 tbsp dark soy sauce
1/2 tsp salt
1 tsp sugar
2 tbsps stock
1/2 tsp sesame oil

生粉 1 1/2 茶匙
水 4 湯匙
* 拌勻

1 1/2 tsps caltrop starch
4 tbsps water
* mixed well

做法
Method

1. 雲耳及冬菇浸洗，剪去硬蒂；鮑魚菇及小粟米切開兩半；菜心摘段，洗淨；甘筍切成花狀。

2. 菜心用油鹽炒熟；草菇剞成十字狀，所有材料飛水，備用。

3. 燒熱油 3 湯匙，加入冬菇、鮑魚菇、草菇、蘑菇、雲耳、小粟米及生筋炒勻，下調味料拌勻至香味散發，最後加入菜心、甘筍片及獻汁略炒，上碟即可。

1. Soak the cloud ear fungus and black mushrooms in water and wash. Cut away the hard stalks with scissors. Cut the oyster mushrooms and baby corn in half. Pick the flowering Chinese cabbage into sections and rinse. Cut the carrot into floral shape.

2. Stir-fry the flowering Chinese cabbage with oil and salt until done. Score the straw mushrooms in crisscross. Scald all the ingredients. Set aside.

3. Heat 3 tbsps of oil. Stir-fry the black mushrooms, oyster mushrooms, straw mushrooms, button mushrooms, cloud ear fungus, baby corn and gluten balls. Add the seasoning. Stir-fry until fragrant. Finally put in the flowering Chinese cabbage, carrot and thickening glaze. Roughly stir-fry. Serve.

蛋白火鴨絲卷
Egg White Rolls with Roast Duck

材料
Ingredients

燒鴨肉 4 兩
雞蛋白 8 個
皮蛋 2 個
酸子薑 1 兩
青瓜半條

150 g roast duck meat
8 egg whites
2 preserved duck eggs
38 g pickled young ginger
1/2 cucumber

小秘訣 Tips

用慢火煎蛋白一面即可，別反轉再煎，蛋白才滑嫩。

To make silky egg white skin, fry only one side of the egg white over low heat. It is no need to fry the other side.

做法
Method

1. 蛋白與半茶匙鹽輕輕拌勻，舀入有少許油的平底鑊內，煎成薄蛋白皮。
2. 燒鴨肉、皮蛋及子薑切成 2 吋長幼絲，拌入少許麻油，備用。
3. 將一塊薄蛋白皮切成四份，包入上述餡料，捲成長包狀，排於碟上，伴青瓜片享用。

1. Gently mix the egg whites with 1/2 tsp of salt. Ladle in a pan containing a little oil. Fry into egg white skin.
2. Finely cut the roast duck, preserved duck eggs and pickled young ginger into shreds of 2-inch long. Mix in a little sesame oil. Set aside.
3. Cut a sheet of the egg white skin into quarters. Wrap the above stuffing. Roll into a cylinder. Arrange on a plate. Serve with the cucumber.

焗薯茸肉卷
Baked Mashed Potato and Pork Rolls

材料
Ingredients

豬肉碎 12 兩
洋葱 1 個（小型，剁碎）
馬鈴薯 2 個（約半斤）
牛油 2 湯匙
麵包糠 1 杯

450 g minced pork
1 small onion (finely chopped)
2 potatoes (about 300 g)
2 tbsps butter
1 cup breadcrumbs

薯茸調味料
Seasoning for mashed potato

鹽及糖各半茶匙
蛋液半個
鮮忌廉或鮮奶 1 湯匙
胡椒粉少許
芫茜 2 湯匙

1/2 tsp salt
1/2 tsp sugar
1/2 egg wash
1 tbsp cream or milk
ground white pepper
2 tbsps coriander

肉碎調味料
Seasoning for minced pork

鹽、糖及紹酒各半茶匙
生抽 1 茶匙
蛋液半個
鮮忌廉或鮮奶 1 湯匙
咖喱粉 1 1/2 湯匙
胡椒粉少許

1/2 tsp salt
1/2 tsp sugar
1/2 tsp Shaoxing wine
1 tsp light soy sauce
1/2 egg wash
1 tbsp cream or milk
1 1/2 tbsps curry powder
ground white pepper

小秘訣 Tips

錫紙必須均勻地掃上牛油，以免烤烘後肉醬黏着錫紙。

The aluminum foil must be evenly brushed with butter; otherwise, the pork will stick to the aluminum foil after baked.

做法
Method

1. 馬鈴薯洗淨，放入沸水內炝熟至腍，取出，去皮，壓成薯茸，下調味料拌勻，備用。

2. 取半份肉碎，下油鑊爆炒至熟，待涼；燒熱牛油1湯匙，下洋葱茸炒香，待涼。

3. 將已炒的肉碎及洋葱拌入半份未煮肉碎內，下調味料拌勻。

4. 錫紙掃上牛油，鋪上麵包糠成長方形狀，再均勻地放上肉醬，中央放上薯茸，輕輕將錫紙捲起定型，做成薯茸餡肉卷，再揭開錫紙，灑上麵包糠，捲緊。

5. 放入預熱焗爐，用250℃焗約半小時，至肉餡全熟及散出香味，切件享用。

1. Rinse the potatoes. Put in the boiling water and cook until tender. Remove the skin. Mash up. Mix in the seasoning. Set aside.

2. Take half portion of the minced pork. Stir-fry in oil until done. Let it cool down. Heat 1 tbsp of butter. Stir-fry the onion until fragrant. Let it cool down.

3. Mix the stir-fried pork and onion with the rest portion of the uncooked pork. Add the seasoning. Mix well.

4. Brush butter on aluminum foil. Lay the breadcrumbs on top and spread into a rectangular shape. Put the minced pork mixture evenly on top. Place the mashed potato in the middle. Gently roll the aluminum foil, making a mashed potato and pork roll. Open the aluminum foil and sprinkle with the breadcrumbs. Roll tightly.

5. Put into a preheated oven. Bake at 250℃ for about 1/2 hour, or until the pork is fully done and smells fragrant. Cut into pieces. Serve.

葉 胡 影 儀 女 士
Mrs. Yip Woo Ying Yee

被學生稱為「鄰家的媽媽」的葉太，心靈手巧、廚藝精湛、熱愛烹飪，尤其炮製家常小菜更是拿手絕活。

廚藝顯真情 Cooking with true feeling

20 年以來，葉太憑着對烹飪的熱誠，勇於實踐天賦，成為專長的職業，在聖公會聖匠堂教授再培訓課程及陪月課程；於屯門明愛僱員再培訓中心、社會福利署長沙灣中心、香港遊樂場協會等多間社區中心擔任烹飪班導師。

葉太授課時講解生動，以簡易的做法炮製特色美食，透過食材的運用及配搭，讓學生掌握材料的特點，體驗烹飪的樂趣，桃李滿門。

菜餚會摯友 Great delicacies for friends

葉太性情和善，樂意與別人分享美食，經常在家款待良朋摯友，為大家炮製各款令人嘖嘖稱讚、色香味俱全的菜餚。友人邊品嘗美餚，邊輕談淺酌，滿室歡聲笑語。

Respected by students as "Mother of the neighbourhood" for her skilful and excellent cookery, Mrs. Yip has a passion for cooking with a flair for home dishes.

In the past 20 years, Mrs. Yip has developed her cooking fervour into a teaching profession. She tutors the art of cooking in employees retraining and post-natal care programmes at the Sheng Kung Hui Holy Carpenter Church; and cookery courses in various community centres such as the Caritas Tuen Mun Employees Retraining Centre, Cheung Sha Wan Community Centre of the Social Welfare Department, and the Hong Kong Playground Association

Mrs. Yip let her many students enjoy the pleasure of cooking. By giving a colourful account of cookery, she teaches them how to use simple techniques to make special and favourite delicacies, and how to make use of the characteristics of different materials to make delicious cuisine through the complementary use of various ingredients.

Nice and kind, Mrs. Yip is happy to share flavourful food with others at home. She always receives applause for her amazing and savoury dishes. The gleeful laughter fills the house where guests taste the great delicacies while chatting.

100道靚餸

YUMMY DELICACIES

作者	Author
葉胡影儀	Yip Woo Ying Yee
策劃/編輯	Project Editor
	Karen Kan
攝影	Photographer
	Imagine Union
美術統籌	Art Direction & Design
	Ami
美術設計	Design
	Man
出版者	Publisher
	Forms Kitchen
香港鰂魚涌英皇道1065號	Room 1305, Eastern Centre, 1065 King's Road,
東達中心1305室	Quarry Bay, Hong Kong.
電話	Tel 2564 7511
傳真	Fax 2565 5539
電郵	Email info@wanlibk.com
網址	Web Site http://www.wanlibk.com
	http://www.facebook.com/wanlibk
發行者	Distributor
香港聯合書刊物流有限公司	SUP Publishing Logistics (HK) Ltd.
香港新界大埔汀麗路36號	3/F., C&C Building, 36 Ting Lai Road,
中華商務印刷大廈3字樓	Tai Po, N.T., Hong Kong
電話	Tel 2150 2100
傳真	Fax 2407 3062
電郵	Email info@suplogistics.com.hk
承印者	Printer
中華商務彩色印刷有限公司	C&C Offset Printing Co., Ltd.
出版日期	Publishing Date
二〇一三年九月第一次印刷	First print in September 2013
二〇一九年九月第五次印刷	Fifth print in September 2019